Application and Characterization of Rubber Materials

Edited by Gülşen Akın Evingür and Önder Pekcan

Published in London, United Kingdom

Application and Characterization of Rubber Materials
http://dx.doi.org/10.5772/intechopen.102183
Edited by Gülşen Akın Evingür and Önder Pekcan

Contributors
Ntalane S. Seroka, Raymond Taziwa, Lindiwe Khotseng, Mounir Kassmi, Buddhima Rupasinghe, Azemi bin Samsuri, Hammat H. Valiev, Yury V. Kornev, Alexander N. Vlasov, Yulia N. Karnet, Alla A. Kornilova, Nikolay A. Semenov, Józef T. Tadeusz Haponiuk, Justyna Kucińska Lipka, Jacek Rafał Kędzia, Anna Maria Sitko

First published in London, United Kingdom, 2023 by IntechOpen
IntechOpen is the global imprint of INTECHOPEN LIMITED, registered in England and Wales, registration number: 11086078, 5 Princes Gate Court, London, SW7 2QJ, United Kingdom

British Library Cataloguing-in-Publication Data
A catalogue record for this book is available from the British Library

Additional hard and PDF copies can be obtained from orders@intechopen.com

Application and Characterization of Rubber Materials
Edited by Gülşen Akın Evingür and Önder Pekcan
p. cm.
Print ISBN 978-1-80355-606-2
Online ISBN 978-1-80355-607-9
eBook (PDF) ISBN 978-1-80355-608-6

Meet the editors

Dr. Gülşen Akın Evingür is an associate professor at Pîrî Reis University, Istanbul, Turkey. She completed her BSc at Yildiz Technical University, Istanbul, Turkey, and her Ph.D. at Istanbul Technical University in 1996 and 2011, respectively. She has been engaged in various academic studies in the fields of composite gels and their optical, electrical, and mechanical properties. She has authored more than 55 journal articles and 100 proceedings in national and international journals. She has edited three books and authored six book chapters and participated in both national and international projects.

Prof. Pekcan received his MSc in Physics at the University of Chicago, USA in 1971. He obtained his Ph.D. in Physics from the University of Wyoming, USA in 1974. He was a visiting scientist at the Abdus Salam International Centre for Theoretical Physics (ICTP), Italy, and the Technical University of Gdansk, Poland. He was also a visiting professor in the Department of Chemistry, at the University of Toronto, Canada. Dr. Pekcan was appointed a full professor in the Department of Physics, at Istanbul Technical University, where he worked until 2005. Currently, he is a professor in the Department of Bioinformatics and Genetics, at Kadir Has University, Turkey. He has been a member of the Science Academy since 2012. He has more than 380 journal articles, 28 book chapters, and 10 projects to his credit.

Contents

Preface

Rubber materials are used across a huge range of domestic and industrial applications. There are ten common types of rubber, including natural rubber, styrene-butadiene rubber, butyl, nitrile, silicone, polyurethane, hydrogenated nitrile, and so on. This book discusses several different types of rubber materials and their mechanical, optical, acoustic, and kinetic properties.

This book includes six chapters. Chapter 1 by Buddhima Rupasinghe discusses the recycling of silicone-based materials. Chapter 2 by Azemi Samsuri discusses recycled carbon black based on styrene-butadiene rubber, natural rubber, and nitrile rubber compounds. Chapter 3 by Hammat Valiev et al. describes the influence of composition and structure on the properties of elastomeric composites with silicon dioxide fillers. Chapter 4 by Ntalane Sello Seroka et al. discusses sugar cane bagasse ash, an agricultural residue with potential rubber filler applications. Chapter 5 by Józef Haponiuk et al. examines the origin, specifications, and applications of natural rubber latex. Finally, Chapter 6 by Mounir Kassmi presents the characterization of hydrogenated amorphous silicon using infrared spectroscopy and ellipsometry measurements.

We would like to thank all the authors for their excellent contributions. We would also like to thank the staff at IntechOpen, especially Author Service Manager Maja Bozicevic for her effective editing and support during the production of this book.

Gülşen Akın Evingür
Faculty of Engineering,
Pîrî Reis University,
Tuzla, İstanbul, Turkey

Önder Pekcan
Faculty of Engineering and Natural Sciences,
Kadir Has University,
Cibali, İstanbul, Turkey

Chapter 1

Recycling Silicone-Based Materials: An Overview of Methods

Buddhima Rupasinghe

Abstract

Since the early 1800s, siloxane has been an industrial staple due to its remarkable structure, but even though there are many benefits for using siloxanes, there are significant environmental implications, one of which being the lack of recyclability. As the first step to polymerization or the depolymerization of polymers, the scission of the silicone bond is essential. While condition-specific reactions investigating what triggers polymerization have been extensively studied, traditional synthesis methods are unfortunately not ideal due to their high cost and detrimental release of greenhouse gases. Since the 1950s, several studies have related to rupturing the siloxane bond, including hydrolysis, catalytic depolymerization, thermal depolymerization, and radical extractions. This work has resulted in new polymers, cyclics, and monomeric silanes. However, only a few studies have focused on how to build new silicone-based materials from the primary siloxane cyclic forms. Thus, more investigation into better methods for recycling siloxanes is needed. This chapter summarizes and categorizes the published data on the degradation and depolymerization of polysiloxanes based on their reaction temperature up to July 2021.

Keywords: siloxane degradation, silicones, polymer recycling, polymer depolymerization, PDMS

1. Introduction

There is evidence that silicone-based materials have been used for tool making since the Stone Age. During the 27 BC and 14 AD, from Sidon and Babylon to the Roman Empire Ages, ancient people used silicone-based materials for glass manufacturing [1]. After the discovery of the silicon element in the early 1800s by J.J. von Berzelius [2], silicone chemistry began being explored due to the findings of James Crafts and Charles Friedel with the synthesis of the first silicon-organic compound in 1863 and Ladenburg's development of the first polymeric siloxane in 1872 [3, 4]. During the early 1900's Fredric Kipping's work on the synthesis of organosilicon, Eugene Rochow's and James Hyde's work on the direct synthesis of silicon-based materials expanded silicone-based materials into commercial materials [5, 6].

Today, siloxane polymers or silicone-based materials are an important industrial commodity based on the structure $-(R_2\text{-Si-O})-$. **Figure 1** demonstrates how R can be a series of hydride, alkyl, vinyl, and aryl functionalities [7], and the IUPAC named the repeating unit "siloxane" [8]. Polydimethylsiloxane (PDMS) silicones, mainly linear

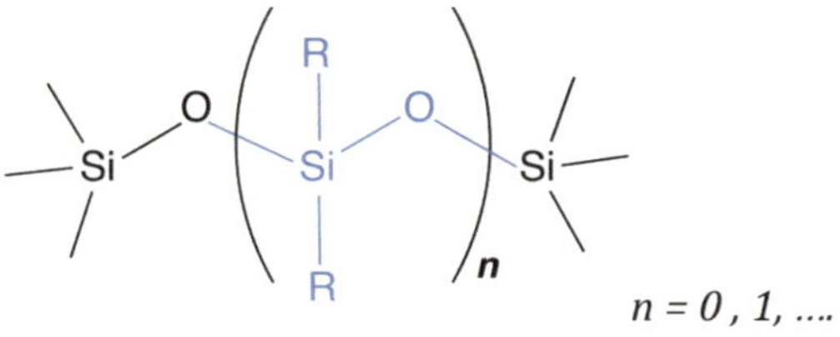

Figure 1.
Trimethylsilyl-terminated polydimethylsiloxane.

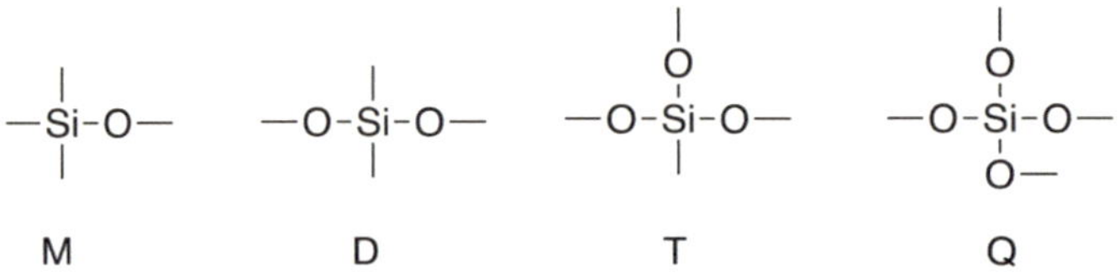

Figure 2.
Siloxane-bonding motifs.

polymeric liquids even for larger n-values, are the most common example. **Figure 2** highlights how a silicon atom connected to one atom is known as M, and structures D, T, and Q are similarly defined. The combination of these units creates different valuable materials such as resins, oils, and elastomers [9].

The hybrid nature of siloxanes with organic and inorganic components leads to unique properties not possible with other materials [10]. For example, their ability to modify their structure allows them to morph into different chemical properties such as hydrophobicity. Additionally, their properties allow for easy alteration of their melting and refractive index, boiling points, viscosity, and density [11]. Likewise through different R-groups, processing conditions, and polymer lengths, the thermal stability properties of these materials can be modified [12] and their non-toxic nature enables a variety of biological applications [13]. As a result, polysiloxanes have become widely used exceptional polymers due to their excellent physical [10], chemical [14], and mechanical properties [15], which are important in medical, cosmetic, aerospace, and other high-tech industries [10]. To combine these superior properties into an advanced system, commercial performance silicone materials are formulated as complex multi-component systems leading to elastomers and resins (**Figure 3**) [17, 18].

By using different types of additives, catalysts, and filler materials, it is possible to create a multi-component 3D network system, but it is important to note that the initial polymer structure and functionality are essential for final properties. For example, the formation of the cyclic distribution after depolymerization would significantly differ if the same PDMS system is different in the network system [19]. Likewise, if the network system consists of a copolymer of PDMS and polyphenylmethylsiloxanes (PMPS) or PDMS and nanocomposites, the decomposition pattern will differ based on their network structure [20]. Also, based on the addition of reinforcing agents, the strength or the mechanical properties and the onset temperature of degradation of the silicone matrix are changed [21]. There are different reinforcing agents used in PDMS systems. For instance, silicon metal oxides [15], carbon black [22], montmorillonite nanocomposite [23], silicates [24], fume silica [25], nano-silica [26], and carbon & SiC-based fibers [27]. Sometimes, filler is added to the system to affect other properties. For example, carbon black is added to improve electrical

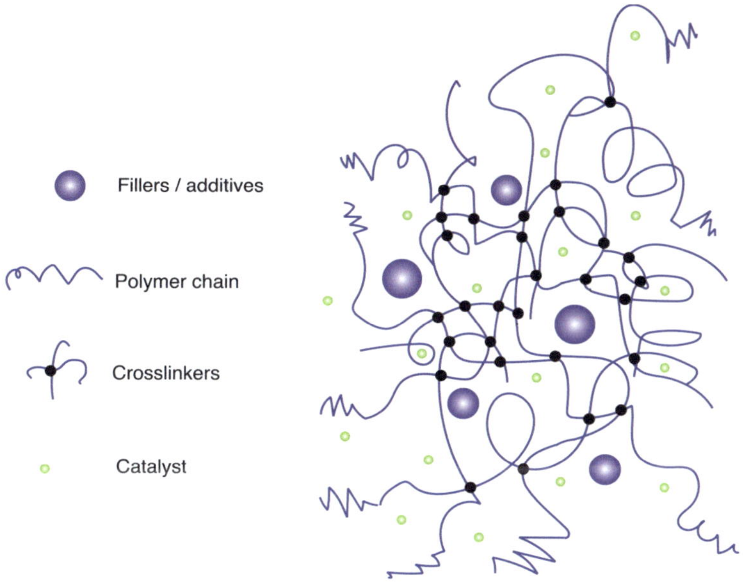

Figure 3.
Complex multicomponent silicone elastomer [16].

conductivity, titanium dioxide improves dielectric constant, and barium sulfate improves radiopacity. Because of these advantages, polysiloxanes are widely used in an immense number of industrial applications [28] varying from simple baking and measuring cups [29] to surface modifiers (elastomers/sealants, antifoaming agents, personal care products, surfactants, coatings, insulations) [30], biomedical [31], microporous organic and inorganic materials [32], and aerospace and other high-tech industries [33].

The majority of the technological products are made from PDMS and to a smaller extent PMPS. As a result, they are considered as the classical products of the silicones. Due to high- and low-temperature resistance, suitable siloxanes can be heated up to 200°C for a year without degradation and even for short terms at 450°C [34]. This, as well as a high degree of chemical inertness [35], makes them especially difficult to degrade and recycle controllably due to the strength of the Si-O bond, which is ~450 kJ/mol [36]. Another disadvantage of this material is synthesizing cost. Synthesis of the Silicon metal involves exclusive process that relies on carbothermal reduction by heating the silica with carbon at >2000°C. In addition to the high cost, this process also unfortunately releases a mol of CO_2 for each mol of silicon metal produced [37] and needs to further functionalize to form different monomeric units for advanced polymerization [9]. Due to the environmental and financial costs, methods to break down silicon-based materials have been explored extensively through hundreds of papers; however, so far only a few studies have considered recycling.

1.1 Polysiloxanes, -Si-0- bond breaking approaches

The deconstruction of polymeric siloxane materials depends on the structure of the Si-O bond and the combination of the M, D, T, and Q units. Notably, this deconstruction often makes product identification challenging to determine. As the makeup of the siloxane bond (Si-O) is unique, it formed σ-bond using s and p electrons of the silicon and p electrons of oxygen and π-bond using unshared p electrons from the oxygen and 3d orbitals from silicon. According to the the Schomaker-Stevenson rule, siloxane bond length 1.64 Å, is lower than its theoretical length of 1.77 Å, due to partial π- and σ-bond character [38]. Additionally, the higher Si-P bond angle compared to the C-P bond angle is unusual, as seen in in the bond angle of dimethyl ether at 111.5° [39] when it increases to 144.1° in disiloxane, partially due to the larger radii of Si and partial delocalization of the electron lone pairs of oxygen [40]. These significant features can be explained by the (p-d) σ-and π-bonding nature and the slightly higher electronegativity of the silicon atom compared with the carbon [41]. While the formation of four tetrahedral σ-bonds is similar to carbon, free 3d orbitals corresponding to the M shell are allowed since silicone is in the third period, resulting in six as the maximum coordination of silicon [42]. The silicon atom can form a wide variety of σ and π bonds with other elements due to its angular functions of the d orbitals. These unoccupied 3d orbitals can act as an acceptor in the donor-acceptor (p-d) π bond. Also, the concept of "one angle nonbonded radii" explains the large bond angles and short Si-O bond [43, 44].

The remarkable properties of silicone materials are due to these unique properties and inorganic hybrid nature [10]. The methods discussed below [41] highlight how even though siloxanes are chemically and thermally stable polymers up to 350°C [45], the presence of a strong acid or a base with moisture can easily break the siloxane bond.

Even though degradation of polysiloxane primarily usually occurs along the Si-O backbone, it can also take place on the substituent atoms or groups (i.e., Si-C bonds), leading to reduced molecular weights. **Figure 4** summarizes the four general processes of siloxane breakdown based on their reaction temperature. Catalytic or non-catalytically driven, depolymerization can be from strong acids or bases (25–110°C), nucleophiles or electrophiles use with catalysts (110–350°C), thermal depolymerization (over 350–500°C), and radical scission & calcination (over 500°C) [46]. Lewis or Bronsted acids and bases [12], organo-metallic catalysts [47], alcohols [48], and radiation [49] are the main catalysts or catalytic activity involved in the depolymerization of siloxanes. As the summary of methods in **Figure 4** shows [18], the siloxane degradation mainly depends on temperature, oxidative or reductive atmospheres, sample mass, and geometry, and the network architecture [50].

Even though the disruption of the Si-O bond from one location is enough to initiate degradation in linear siloxane systems, when the siloxane polymers have a

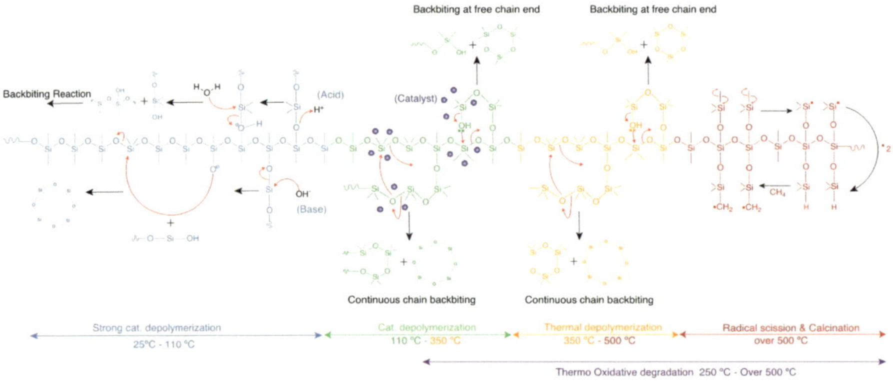

Figure 4.
Mechanisms for the breakdown of siloxanes by various processes.

3D network structure, for degradation, it is imperative to rupture from two or more points [51]. Backbiting reaction mechanisms can explain hydrolysis and catalytic depolymerization reactions [52] and these are preferable in the range of 110–250°C and, in some instances, at room temperature [53]. Additionally, backbiting reactions, radical scission, and calcination reactions can explain thermal depolymerizations. Cyclic, hexamethylcyclotrisiloxane (D_3), octamethylcyclotetrasiloxane (D_4), decamethylcyclopentasiloxane (D_5), and dodecamethylcyclohexasiloxane (D_6) are the prevailing thermodynamic products of siloxane degradation, and enrichment in certain produces is possible depending on the reaction environment and degradation process. For example, the D_3 cyclic form is prominent in the thermal degradation process because products have more variability in lower temperature breakdowns, and the oxidative atmospheric environment is better than the inert atmospheric environment for siloxanes' degradation process because it helps the radical cleavage of the siloxane bond above 250°C [54]. Moreover, siloxane mass [55], geometry [56], and active chain ends [19] contributes to the depolymerization of siloxane.

2. Low-temperature depolymerization, (25 to 110°C)

2.1 Chemical methods for the depolymerization

Particularly in dimethyl derivatives, siloxane bonds are sterically unhindered and extremely polar due to their π-bond and σ-bond character [44]. Consequently, protons can easily protonate the Si-O bond when there is a strong electrophile or nucleophile present. Accordingly, if water or the nucleophile does a nucleophilic attack on the silicon atom, it will form two silanols that can be further rearranged from backbiting reactions to form cyclic derivatives or free silanols [46]. It is noteworthy that either direct substitution or the d-orbitals can cause this depolymerization, and comprehensive studies of these processes are still needed. Common electrophiles for the depolymerization of polysiloxanes are sulfuric acid, triflic acid, and hydrofluoric acid (H.F.] [57]. Some groups have tried strong nucleophiles such as amines, organotin carboxylates, octanoate catalysts, alcohols, and halides [58]. These studies showed that strong electrophile or nucleophiles can breakdown the siloxane bond and can form series of cyclic forms under lower temperature [59].

2.2 Irradiation methods

Apart from the chemically driven bond breaking strategies, there are a limited number of studies that have focused on the irradiation methods to break the -Si-O- bond. Some studies found that PDMS can be degraded at ambient temperature ~ 30°C, using γ-irradiation under the dose rate of 0.35 Mrads/hr. of ^{60}Co can form D_4, D_5, and a little amount of D_6 as their major products. Furthermore, irradiation results were compared with the thermal decomposition for the same material, and it was observed strained bonds are not favorable to form highly excited siliconium ion from the γ-irradiation. As a result, D_4 was observed as the major cyclic form, about ~50%, D_5 ~ 45% by weight and no D_3 form from γ-irradiation. On the other hand in thermal decomposition, D_3 was observed as the major cyclic form about, ~ 60%, and D_4 about ~30% by weight [49]. Based on the dose range, time of the exposure and exposure temperature determined the effectiveness of the radiation depolymerization [60]. Moreover, Xiao et al. studied the degradation of siloxane polymers

Figure 5.
Depolymerization of Wacker Elastosil to cyclic siloxanes in a tetrabutylammonium fluoride THF bath [63].

under direct proton exposure resulting in D_3 structure formation [61]. Additionally, a few studies have used ultrasonic methods to instill depolymerization. The recycling of silica-filled rubber was used to depolymerize and repolymerize it back to polymers using ultrasonic energy at 170°C [62].

2.3 Recycling methods to form siloxane polymers at lower temperatures

It is noteworthy that the above-mentioned studies all focus on the breakage of the -Si-O- bond at lower temperatures. However, they did not focus on repolymerization of the monomeric units that they made due to the separation issues. The main reason for this is that cyclic forms are kinetically stable products. As a result, with time, those different cyclic forms together convert to oligomers to form the thermodynamically stable products.

Recently, Rupasinghe et al. explored an efficient room temperature and low mol% (0.5) catalytic fluoride ($Bu_4N^+F^-$) technique to depolymerize rapidly nearly any silicone-based polymers, elastomers, and resins in the presence of high-swell organic solvents such as Tetrahydrofuran THF (**Figure 5**) [63]. The main products observed were an equilibrium mixture of D_4, D_5, and D_6 verified by ^{29}Si NMR and Gas chromatography–mass spectrometry (GCMS). Notably, if D_5 was used as the starting material, the same products were observed. Furthermore, while complex systems show intricate products alongside distinct cyclics, they observed that silicone-rich systems display the best conversions and the best quantity of distinguishable cyclic forms. By quenching the fluoride ions, they could lock the depolymerization products and enable the collection of the cyclics for repolymerization. They found that if the catalyst is not quenched, cyclics forms can further polymerize to form series of oligomers. This process has strong potential for large-scale industrial processing because it requires minimum energy and lower chemicals. Also, this can depolymerize any commercial silicone-based polymer regardless of additives.

3. Mid-temperature depolymerization, 110 to 350°C

Si-O bond scissions arising along the main chain propel catalytic thermal depolymerization mechanisms [64]. Si-C (78 kcal/mol) bond dissociation energy is lower compared to Si-O bond dissociation energy (108 kcal/mol) [65]. However, Si-O bond scission occurred over Si-C due to the empty silicon and oxygen d-orbitals, which also contributes to the chain flexibility.

The compulsory temperature is lower than the thermal process since this process includes the use of catalysts, and products are most often formed by backbiting reactions (**Figure 4**). Trimethylaluminum and $AlCl_3$, are common electrophiles for depolymerization of siloxanes by catalytic thermal methods, and common nucleophiles for the depolymerization of siloxanes are alkali metal hydroxides; ammonium or phosphonium bases; alkali metal hydroxides; alkali metal halides; and alcohols. It was noted that in order to break the silicone polymer and make cyclic forms, these reactants need heating support. Also, the dominant product is D_4 from the cyclic forms formed ranging from D_3 to D_8 [48, 66].

3.1 Recycling methods to form siloxane polymers at mid-temperatures

Enthaler et al. studied the depolymerization of polysiloxanes using fatty acid anhydride in the presence of catalytic amount of iron salts. They found out when the optimum conditions for silicone oil M100 with fatty acid anhydride (1:3 mol ratio), 5 mol% iron (III) chloride were heated for 24 hours at 200°C, they would receive four different anhydrated depolymerized products. Afterward, they added water and heated for 24 h at 100°C and observed the cyclic formation of D_4, D_5 and hydrolysis of the acid anhydride to the acid. Then, they mentioned that they can use the cyclic forms to build long chain silicones. However, their study showed that the depolymerization was difficult when they used complex or higher-molecular-weight silicone polymers [67].

4. High-temperature thermal degradation, over 350°C

As the primary source for prediction of thermal stability of polysiloxanes, high-temperature thermal degradation has been the most studied depolymerization method. Thermally propelled intramolecular exchange processes can decompose polysiloxanes and favor cyclic siloxane forms [68]. Thermal degradation can be described by two opposing mechanisms. First, the catalytic depolymerization method can be explained from the molecular mechanism, with the difference being that the thermal process does not utilize additional catalysts. Second, the homolytic cleavage of Si-C bonds causes the radical mechanism [69]. As it evolves methane through hydrogen abstraction, these mechanisms are theoretically and experimentally observable in higher temperatures [70]. To decrease polymer flexibility and extinguish the formation of cyclic depolymerization derivatives, crosslinking of macromolecular radicals can be utilized [71].

A backbiting mechanism drives the uncatalyzed degradation of PDMS and PMPS and form cyclic forms at 350–400°C. The main thermodynamic product of this high-temperature process is cyclic D_3, yielding 30–80% with a tailing distribution of cyclic structures from D_4 to D_{24} [72]. These reactions are the thermodynamically necessary reverse of the ring-opening polymerization reactions. In other words, unmodified linear PDMS in the absence of catalyst residues or impurities is stable up to 350°C [73]. Thermodynamically, polymerization of cyclic silicone to form linear polysiloxane occurs at ~110°C [74].

5. Discussion and conclusion

As highlighted in this chapter, the degradation of siloxane-based materials has been extensively studied. Since the 1950s, there have been several publications related to rupturing the siloxane bond, including hydrolysis, catalytic depolymerization, thermal depolymerization, and radical extractions. This work has resulted in new polymers, cyclics, and monomeric silanes. It is noted that siloxane polymer degradation depends on various factors. First, structural architecture (filler type/level, crosslink density, chemical substitution of the polymer backbone, inter-crosslink chain length, degree of free chain ends, and the active chain ends) and the degradation method (oxidative or reductive atmospheres) mainly determine the thermal degradation behavior of the siloxane polymer. Second, the depolymerization source (chemical or physical) has its own important parameters to increase the efficiency. Also, for the polysiloxanes degradation, the most common and deeply analyzed mechanism is thermal degradation.

In the chemical approaches, various types of nucleophiles and electrophiles have been analyzed for the depolymerization efficiency throughout the years. Based on their strength, the depolymerization effectiveness varies. For instance, the strong nucleophile can break the -Si-O- bond at room temperature where the weaker nucleophiles can break with heat. Likewise, F^- ions and the amines behave better with the depolymerization of polysiloxanes.

It is notable that the mechanism for the irradiation process has not yet been researched extensively. Free radical processes are achievable if the ionic character is low in the siloxane skeleton because of the irradiation. More significant, however, is that kinetic factors illustrate only a slight role in several heteroatom equilibrations as thermodynamic factors become more crucial with higher temperatures. Consequently, statistical considerations explain the formation of the cyclic products due to the ring-chain equilibration, including the concentration of any cyclic species in equilibrium. With its open-chain homolog, cyclic form is evaluated by considering the probability of ring closure, or equilibrations are considered as scrambling viewpoints. High energy could be another potential explanation, as H can be initiated by methyl groups forming H and Si-CH_3 radicals. Additionally, C-Si bonds can be broken, resulting in the formation of methyl and main chain radicals.

Most studies were focused on the degradation or depolymerization of the polysiloxanes. However, as economic and environmental concerns increase, more focus on the development of recycling and reuse methods is needed. Recently, Enthaler [48] Laine [75] and Rupasinghe & Furgal [53] have focused on developing recycling methods. Their studies have explored the reuse of siloxanes either directly or by using recasting methods with the reformation of starting materials to repolymerize. The possibility of reusing siloxanes is promising, and we suspect the field will transit in this direction so that the energy-intensive processes used to create siloxanes are not wasted. As promising as these initial studies have been, however, further development is needed to “green up” the catalytic processes. Specifically, the use of solvent-free or more earth-friendly solvents can be explored as potential drivers of catalytic processes. Also, the use of other alternative energy sources, such as light, to cause on-demand depolymerization could also trigger the catalytic breakdown of siloxane elastomers and resins in commercial use.

Acknowledgements

The authors thank the Bowling Green State University Building Strength Program and the Faculty Startup Program for Funding.

Conflict of interest

The author declares no conflict of interest.

Author details

Buddhima Rupasinghe
Charles River Laboratories International, Inc., Mattawan, MI, United States

*Address all correspondence to: buddhima.rupasinghe@crl.com

References

[1] Fascinating Silicones Proven Innovation | Dow Inc. [Internet]. [cited 2022 Jun 29]. Available from: https://www.dow.com/en-us/fascinating-silicones/fascinating-silicones-proven-innovation.html

[2] Berzelius JJ. Untersuchungen über die Flussspathsäure und deren merkwürdigsten Verbindungen. Annalen der Physik und der physikalischen Chemie. 1824;**78**(10):113-150

[3] Friedel C, Crafts JM. Ueber einige neue organische Verbindungen des Siliciums und das Atomgewicht dieses Elementes. Annalen der Chemie und Pharmacie. 1863;**127**(1):28-32

[4] Ladenburg A. Ueber die Reductionsproducte des Kieselsäureäthers und deren Derivate. Annalen der Chemie und Pharmacie. 1872;**164**(2):300-332

[5] Thomas NR. Frederic Stanley kipping—Pioneer in silicon chemistry: His life & legacy. SILICON. 2010;**2**(4):187-193

[6] Rochow EG, Gilliam WF. Polymeric methyl silicon oxides 1. Journal of the American Chemical Society. 1941;**63**(3):798-800

[7] Clarson SJ, Fitzgerald JJ, Owen MJ, Smith SD. Silicones and Silicone-Modified Materials, Copyright, Foreword. In: Silicones and silicone-modified materials. 2000;729:i–v. https://doi.org/10.1021/bk-2000-0729.fw001

[8] LeVier R. What is silicone? Journal of Clinical Epidemiology. 1995;**48**(4):513-517 [Internet]. Available from: https://linkinghub.elsevier.com/retrieve/pii/0895435694002077

[9] Pawlenko S. Organosilicon Chemistry. Walter de Gruyter. B: DE GRUYTER; 1986. https://doi.org/10.1515/9783110862386

[10] Moretto HH, Schulze M, Wagner G. Silicones. In: Ullmann's Encyclopedia of Industrial Chemistry. Weinheim, Germany: Wiley-VCH Verlag GmbH & Co. KGaA; 2000

[11] Mark JE. Physical properties of polymers handbook. In: Mark JE, editor. Dictionary Geotechnical Engineering/Wörterbuch GeoTechnik. New York, NY: Springer New York; 2007. pp. 211-211

[12] Grassie N, Macfarlane IG. The thermal degradation of polysiloxanes. Part 1. Poly(dimethylsiloxane). European Polymer Journal. 1978;**14**(11):875-884

[13] Potter M, Rose NR, Compans RW, Cooper M, Koprowski H, Melchers F, et al. Immunology of Silicones. Vol. 6. Berlin Heidelberg: Springer; 1996. pp. 1-423

[14] Cogan HD, Setterstrom CA. Properties of ethyl silicate. Chemical & Engineering News. 1946;**24**(18):2499-2501

[15] Wen J, Mark JE. Synthesis, structure, and properties of poly(dimethylsiloxane) networks reinforced by in situ-precipitated silica–titania, silica–zirconia, and silica–alumina mixed oxides. Journal of Applied Polymer Science. 1995;**58**(7):1135-1145

[16] Lewicki JP, Maxwell RS. Degradative thermal analysis of engineering silicones. In: Concise Encyclopedia of High Performance Silicones. Hoboken, NJ, USA: John Wiley & Sons, Inc.; 2014. pp. 191-210

[17] Jacobine AF, Nakos ST. Photopolymerizable silicone monomers, oligomers, and resins. In: Radiation Curing. Boston, MA: Springer US; 1992. pp. 181-240

[18] Rupasinghe B, Furgal JC. Degradation of Silicone-based Materials as a Driving Force for Recyclability. Polym Int 2021, No. August. https://doi.org/10.1002/pi.6340

[19] Lewicki JP, Mayer BP, Alviso CT, Maxwell RS. Thermal degradation behavior and product speciation in model poly(dimethylsiloxane) networks. Journal of Inorganic and Organometallic Polymers and Materials. 2012;**22**(3):636-645

[20] Kroonblawd MP, Goldman N, Lewicki JP. Chemical degradation pathways in siloxane polymers following phenyl excitations. Journal of Physical Chemistry B. 2018;**122**(50):12201-12210

[21] Xiang B, Deng Z, Zhang F, Wen N, Lei Y, Liu T, et al. Microcellular silicone rubber foams: The influence of reinforcing agent on cellular morphology and nucleation. Polymer Engineering and Science. 2019;**59**(1):5-14

[22] Zhu H, Wang F, Wang L, Wang Y. Influences of compression cycles on piezocapacitive effect of carbon black filled silicone rubber composite. Sensors and Actuators A: Physical. 2018;**271**:192-200

[23] Yang L, Hu Y, Lu H, Song L. Morphology, thermal, and mechanical properties of flame-retardant silicone rubber/montmorillonite nanocomposites. Journal of Applied Polymer Science. 2006;**99**(6): 3275-3280

[24] Ahn D, Dhinojwala A. Silicone Surface Science. In: Owen MJ, Dvornic PR, editors. Chapter 2: Sum Frequency Generation Vibrational Spectroscopy of Silicone Surfaces & Interfaces. Vol. 4. Dordrecht: Springer Netherlands; 2012. pp. 23-58. (Advances in Silicon Science

[25] Colas A, Curtis J. Silicone biomaterials:History and chemistry. Elsevier Science Publishers BV. 2004;**20**(1):57-60

[26] Fink JK. Liquid Silicone Rubber; Wiley, 2019;**58**. https://doi.org/10.1002/9781119631408

[27] Pielichowski K, Njuguna J. Thermal degradation of polymeric materials. Polimeri: časopis za plastiku i gumu. 2005;**26**:106-106

[28] Arkles B. Look what you can make out of silicones. ChemTech. 1983;**13**:542-555

[29] COPPOCK JBM, Hulse H, Urie A. Silicone Pan Glazes in Bread Production. British Medical Journal. 1949;**2**:579

[30] Gao Y, Liang X, Bao W, Li S, Wu C, Liu Y, et al. Effects of liquids immersion and drying on the surface properties of HTV silicone rubber: Part I-contact angle and surface chemical properties. IEEE Transactions on Dielectrics and Electrical Insulation. 2017;**24**(6):3594-3602

[31] Matisons J. Bio-inspired silicon-based materials. In: Zelisko PM, editor. Earth. Vol. 5. Dordrecht: Springer Netherlands; 2014. p. 424. (Advances in Silicon Science

[32] Shanan Moore Stevens N, Rezac ME. Formation of hybrid organic/inorganic composite membranes via partial pyrolysis of poly(dimethyl siloxane). Chemical Engineering Science. 1998;**53**(9):1699-1711

[33] O'Neill J, Lu J, Dockter R, Kowalewski T. Stretchable, flexible, scalable smart skin sensors for robotic position and force estimation. Sensors. 2018;**18**(4):953

[34] Hunter MJ, Hyde JF, Warrick EL, Fletcher HJ. Organo-silicon polymers. The cyclic dimethyl siloxanes. Journal of the American Chemical Society. 1946;**68**(4):667-672

[35] Graffius G, Bernardoni F, Fadeev AY. Covalent functionalization of silica surface using "inert" poly(dimethylsiloxanes). Langmuir. 2014;**30**(49):14797-14807

[36] Interrante LV, Shen Q. Silicon-containing polymers. In: Jones RG, Ando W, Chojnowski J, editors. Silicon-Containing Polymers. Dordrecht: Springer Netherlands; 2000. pp. 247-321

[37] Abolpour B, Shamsoddini R. Mechanism of reaction of silica and carbon for producing silicon carbide. Progress in Reaction Kinetics and Mechanism. 2020;**45**:146867831989141

[38] Pitzer KS. The nature of the chemical bond and the structure of molecules and crystals: An introduction to modern structural chemistry. Journal of the American Chemical Society. 1960;**82**(15):4121-4121

[39] Kimura K, Kubo M. Structures of dimethyl ether and methyl alcohol. The Journal of Chemical Physics. 1959;**30**(1):151-158

[40] Gillespie RJ, Johnson SA. Study of bond angles and bond lengths in Disiloxane and related molecules in terms of the topology of the electron density and its Laplacian. Inorganic Chemistry. 1997;**36**(14):3031-3039

[41] Bürger H. Anomalies in the structural chemistry of silicon. Angewandte Chemie International Edition in English. 1973;**12**(6):474-486

[42] Craig DP, Zatjli C. D orbitals in compounds of second-row elements. I. SF6. The Journal of Chemical Physics. 1962;**37**(3):601-608

[43] Glidewell C. Some chemical and structural consequences of non-bonded interactions. Inorganica Chim Acta. 1975;**12**(1):219-227

[44] Oberhammer H, Boggs JE. Importance of (p-d).Pi. Bonding in the siloxane bond. Journal of the American Chemical Society. 1980;**102**(24):7241-7244

[45] Lewis FM. The science and Technology of Silicone Rubber. Rubber Chemistry and Technology. 1962;**35**(5):1222-1275

[46] Cypryk M, Apeloig Y. Mechanism of the acid-catalyzed Si—O bond cleavage in siloxanes and Siloxanols. A theoretical study. Organometallics. 2002;**21**(11):2165-2175

[47] Patel M, Skinner AR, Chaudhry A, Billingham NC, Mahieu B. Impact of thermal ageing on the tin catalyst species in room temperature vulcanised polysiloxane rubbers. Polymer Degradation and Stability. 2004;**83**(1):157-161

[48] Enthaler S. Iron-catalyzed depolymerization of polysiloxanes to produce dichlorodimethylsilane, diacetoxydimethylsilane, or dimethoxydimethylsilane. Journal of Applied Polymer Science. 2015;**132**:41287

[49] Tanny GB, st. Pierre LE. Production of cyclic molecules during the γ-irradiation of linear polydimethylsiloxane. Journal of Polymer Science. Part B. 1971;**9**(11):863-868 [Internet]. Available from: https://onlinelibrary.wiley.com/doi/abs/10.1002/pol.1971.110091111

[50] Papkov VS, Bulkin AF, Zhadanov AA, Slonimskii GL,

Andrianov KA. Investigation of the thermal oxidation of polydimethylsiloxanes. Polymer Science USSR. 1977;**19**(4):962-976

[51] Gerber AH, Mclnerney EF. Survey of Inorganic Polymers. Horizons Research Incorporated, Cleveland, OH; 1979:317. Available from: https://apps.dtic.mil/sti/pdfs/ADA305001.pdf

[52] Osthoff RC, Bueche AM, Grubb WT. Chemical stress-relaxation of polydimethylsiloxane elastomers 1. Journal of the American Chemical Society. 1954;**76**(18):4659-4663

[53] Rupasinghe B, Furgal JC. Full circle recycling of Polysiloxanes via room-temperature fluoride-catalyzed Depolymerization to Repolymerizable Cyclics. ACS Appl Polym Mater. 2021;**3**(4):1828-1839

[54] Scala L, Hickam W. Thermal and oxidative degradation of silicones. Industrial and Engineering Chemistry. 1958;**50**(10):1583-1584

[55] Kovalenko VY, Pustyl'nik ML, Bebchuk TS, Muzafarova MN, Zherdev YV, Nedorosol VD, et al. Structure of the monomeric unit of crosslinked polymer and its relation to the thermal stability of polymethylsiloxanes. Polymer Science USSR. 1978;**20**(7):1824-1833

[56] Andrianov KA, Kolchina AG, Varlamova NV, Talanov VN, Yakushkina SE. Thermal oxidative degradationoflinearpolydimethylsiloxanes and polydimethylelementosiloxanes. Bulletin of the Academy of Sciences of the USSR Division of Chemical Science. 1968;**17**(2):349-353

[57] Brook MA, Zhao S, Liu L, Chen Y. Surface etching of silicone elastomers by depolymerization. Canadian Journal of Chemistry. 2012;**90**(1):153-160

[58] Patel M, Skinner AR. Thermal ageing studies on room-temperature vulcanised polysiloxane rubbers. Polymer Degradation and Stability. 2001;**73**(3):399-402

[59] Chang CL, Lin YK. Competition of aminolysis and alcoholysis in nucleophilic cleavage of a model compound for polysiloxane networks. Polymer Degradation and Stability. 2005;**87**(1):207-211

[60] Chinn SC, Alviso CT, Berman ESF, Harvey CA, Maxwell RS, Wilson TS, et al. MQ NMR and SPME analysis of nonlinearity in the degradation of a filled silicone elastomer. The Journal of Physical Chemistry. B. 2010;**114**(30):9729-9736 [Internet]. Available from: https://pubs.acs.org/doi/10.1021/jp1013797

[61] Xiao H, Li C, Yang D, Li X, He S. Optical degradation of polydimethylsiloxane under 150 keV proton exposure. Journal of Applied Polymer Science. 2008;**109**(6):4060-4064 [Internet]. Available from: https://onlinelibrary.wiley.com/doi/10.1002/app.28591

[62] Shim SE, Isayev AI. Ultrasonic devulcanization of precipitated silica-filled silicone rubber. Rubber Chemistry and Technology. 2001;**74**(2):303-316

[63] Rupasinghe B, Furgal JC. Full Circle Recycling of Polysiloxanes via Room-Temperature Fluoride Catalyzed Depolymerization to Repolymerizable Cyclics. ACS Appl Polym Mater 2021;**3**(4);1828-1839. https://doi.org/10.1021/acsapm.0c01406

[64] Camino G, Lomakin SM, Lageard M. Thermal polydimethylsiloxane degradation. Part 2. The degradation mechanisms. Polymer (Guildf). 2002;**43**(7):2011-2015

[65] Assary RS, Curtiss LA, Redfern PC, Zhang Z, Amine K. Computational studies of Polysiloxanes: Oxidation potentials and decomposition reactions. The Journal of Physical Chemistry C. 2011;**115**(24):12216-12223

[66] Kučera M, Jelínek M, Láníková J, Veselý K. Termination in anionic polymerization of octamethylcyclotetrasiloxane. Formation of stable complexes on active sites. Journal of Polymer Science. 1961;**53**(158):311-320

[67] Weidauer M, Heyder B, Woelki D, Tschiersch M, Köhler-Krützfeldt A, Enthaler S. Iron-catalyzed depolymerizations of end-of-life silicones with fatty alcohols. Resource-Efficient Technologies. 2015;**1**(2):73-79. [Internet]. Available from. DOI: 10.1016/j.reffit.2015.09.002

[68] Hall AD, Patel M. Thermal stability of foamed polysiloxane rubbers: Headspace analysis using solid phase microextraction and analysis of solvent extractable material using conventional GC–MS. Polymer Degradation and Stability. 2006;**91**(10):2532-2539

[69] Fang W, Zeng X, Lai X, Li H, Chen W, Zhang Y. Thermal degradation mechanism of addition-cure liquid silicone rubber with urea-containing silane. Thermochimica Acta. 2015;**605**:28-36

[70] Almond MJ, Becerra R, Bowes SJ, Cannady JP, Ogden JS, Young NA, et al. A mechanistic study of the low pressure pyrolysis of linear siloxanes. Physical Chemistry Chemical Physics. 2009;**11**(40):9259

[71] Giuliani JR, Gjersing EL, Chinn SC, Jones TV, Wilson TS, Alviso CT, et al. Thermal degradation in a Trimodal poly(dimethylsiloxane) network studied by 1 H multiple quantum NMR. The Journal of Physical Chemistry. B. 2007;**111**(45):12977-12984

[72] Blazsó M, Garzó G, Andrianov KA, Makarova NN, Chernavski AI, Petrov IM. Thermal decomposition of cyclo-linear methylsiloxane polymers. Journal of Organometallic Chemistry. 1979;**165**(3):273-279

[73] Allan D, Radzinski SC, Tapsak MA, Liggat JJ. The thermal degradation behaviour of a series of siloxane copolymers - a study by thermal volatilisation analysis. SILICON. 2016;**8**(4):553-562

[74] Thomas TH, Kendrick TC. Thermal analysis of polydimethylsiloxanes. I. Thermal degradation in controlled atmospheres. Journal of Polymer Science Part A-2: Polymer Physics. 1969;**7**(3):537-549

[75] Krug DJ, Asuncion MZ, Laine RM. Facile approach to recycling highly cross-linked thermoset silicone resins under ambient conditions. ACS Omega. 2019;**4**(2):3782-3789

Chapter 2

Evaluation of Recycled Carbon Black (r-CB) Based on Styrene Butadiene Rubber, Natural Rubber and Nitrile Rubber Compounds

Azemi bin Samsuri

Abstract

The enormous accumulation of used tyres has long been a threat to the environment. Pyrolysis is a process of chemically decomposing organic materials at elevated temperatures above 430°C in the absence of oxygen. Pyrolysis can be used to reprocess the tyres into fuel gas, oils, solid residue (char) and low-grade carbon black, which cannot be used in tyre manufacture. Rubber compounds containing r-CB were prepared based on Styrene Butadiene Rubber, Natural Rubber and Nitrile Rubber. Specific formulations were selected for each of the rubber. Natural Rubber was mixed with r-CB to produce pre-cured tread compound for retreading, Nitrile Rubber was mixed with r-CB to produce hose and seals compounds, and finally, r-CB was compounded in accordance to the Standard Test Methods for Styrene-Butadiene Rubber Recipe and Evaluation Procedures. The quality and performance of r-CB in these rubber compounds were compared with the commercial virgin carbon black. The results show that r-CB matches the quality of virgin carbon black such as High Abrasion Furnace (N330) and Fast Extrusion Furnace (FEF, N550) in terms of hardness, tensile strength, abrasion resistance and other relevant physical properties. The prospects of r-CB are very bright and promising. The challenges ahead are to maintain high quality of r-CB, to be competitive in cost in relation to virgin carbon black.

Keywords: r-CB, reinforcing black, tensile strength, hardness, pyrolysis

1. Introduction

It has been reported that nearly 300 millions of waste tyres have to be managed in Europe, and over 1 billion end-of-life tyres generated annually, worldwide [1]. As a general rule, truck tyres should last between 25,000 and 50,000 miles. With proper and regular maintenance and attention, the life of a tyre can reach 70,000 miles before the truck tyres have to be replaced. The enormous accumulation of used tyres has long been a threat to the environment. Disposing of used tyres into empty land, rivers, lakes and jungles generates an ideal environment for the breeding of mosquitoes, snakes and rats. Besides that, it can lead to a potential uncontrolled fire hazards.

IntechOpen

Tyre fires can occur easily, burning for months and creating substantial pollution in the air and ground. Deliberate burning in the open air is also forbidden because it breaks air pollution regulations and is no longer allowed. To solve this environmental problem, used tyres have to be reused in a manner that will be environmentally friendly as well as economically viable. The common methods to dispose used tyres are landfill, dumping and stockpiling. Tyres are not desired at landfills, due to their large volumes and 75% void space. Landfill is expensive as a consequence of high price to purchase a piece of land. There are law restrictions in some countries which forbid dumping of used tyres in landfill to avoid potential fire hazards, breeding of mosquitoes, rats and poisonous snakes. One of the promising means of disposing used tyres is by the technique known as pyrolysis. During pyrolysis, used tyres are subjected to heat at elevated temperatures above 430°C in the absence of oxygen. Pyrolysis converts the tyres into fuel gas, oils, solid residue (char) and low-grade carbon black, which cannot be used in tyre manufacture. One of the main issues of recycled carbon black (after this is called r-CB) produced by pyrolysis is concerned with its quality. Although highly reinforcing carbon black grades such as Intermediate Super Abrasion Furnace (ISAF N200 series) and High Abrasion Furnace (HAF N300 series) are used in the tyre compounds, but after pyrolysing, the r-CB produced does not retain the quality of the original virgin carbon black. It was reported that r-CBs reinforcement properties are lower than those of N330 carbon black for a same rubber formula [1].

It is well established that the particle size and chemical surface activity of the carbon black are the determining factor affecting the degree of reinforcement. The aggregates of virgin carbon black are in the range of from 0.09 μm to 0.50 μm. In contrast, r-CB produces aggregates of sizes from ~18 μm – 1 mm [1]. The r-CB composite aggregates composed of the carbon blacks from the tyre, the inorganic ash content, fresh carbon and volatiles [1]. By milling process, the r-CB narrows the particle size distribution to typically ~1–10 μm [1]. The low extent of reinforcement of r-CB is attributed to its big particle size and its low chemical surface activity. In order to improve the quality and consistency of r-CB produced, it is of paramount important to scrutinise and monitor the pyrolysis operation at all stages of the process. Quality control process begins from sorting and segregating of used tyres. To minimise the inconsistency of the quality of r-CB, the first thing to do is not to mix passenger car tyres with truck tyres because the compound formulations of these two tyres are very different. It is also beneficial to segregate tyres according to their respective brands to minimise variations in particular the ash content. The next step is to control the tyre feedstock input, which defines the chemical composition of the r-CB produced. Last but not least, the residence time during pyrolysis and other related controlling parameters. This chapter describes and discusses the work to evaluate the quality of r-CB produced by the local manufacturer Eco Power Sdn Bhd. The main objective is to compare r-CB with the performance of virgin carbon black of the N330, Fine Extrusion Furnace (FEF carbon black N550), General Purpose Furnace (GFP N660) and Semireinforcing Furnace (SRF N760) based on five different compound formulations. The first compound is based on the Standard Test Methods for Styrene-Butadiene Rubber (SBR) Recipe and Evaluation Procedures [2], second is based on black-filled Natural Rubber (NR) standard formulation, third is based on pre-cured tread compound to meet the Malaysian Standard (MS) for retreading tyres MS 1208–2020 (3rd Revision) [3], fourth is based on hose liner compound to meet the British Standard Specification for Rubber Hose and Hose Assemblies for Liquefied Petroleum Gas Lines [4] and lastly, rubber seal (O-ring) compound to meet the American Society for Testing and Materials (ASTM) Standard

Classification System for Rubber Products in Automotive Applications [5]. From this evaluation we can then make some assessments on the prospects, challenges and the future of r-CB to meet its demands in the manufacture of rubber products. Currently, the negative perception of r-CB still exists among the Rubber Compounders in the rubber products' manufacturer. The information laid down in this chapter provides experimental evidence on the quality and performance of r-CB currently produced by the local supplier. With the escalating cost of virgin carbon black, it is timely to consider r-CB as an alternative filler in the manufacture of rubber products since the cost of r-CB is about 40–50% lower than the virgin carbon black. It is very timely that the two big players in the tyre manufacturing, namely Bridgestone and Michelin, are keen to use r-CB in the manufacturing of new tyres because of the environmental benefits of this sustainable material [6]. They have claimed that using r-CB in the new tyre production reduces carbon dioxide (CO_2) emission by up to 85% compared with virgin carbon black [6]. They further envisaged that by substituting just 10% of virgin carbon black by r-CB in making new tyres would reduce CO_2 emissions globally by up 2 million metric tons annually. It is well established that the huge accumulation of waste or used tyres and their disposals are of great concern to us because they pose huge threats to the environments.

2. Evaluation of the physical properties of r-CB based on ASTM D-3191-10 (Reapproved 2014)

Unless otherwise stated, all the work reported and discussed throughout in this chapter used r-CB from Eco Power Sdn Bhd Malaysia. **Table 1** shows the Certificate of Analysis of r-CB produced by LabAlliance Sdn Bhd.

The BET surface area is a measure of the total surface area using the theory of Brunauer, Emmett and Teller (BET) nitrogen gas adsorption [11]. This total surface area includes micropores (where the pore diameter less than 2 nm) existed on the surface of carbon black. In contrast, STSA (Statistical Thickness Surface Area) is a measure of external surface area without considering the area of micropore; thus, STSA is always lower than BET Surface Area. STSA is also claimed as specific surface area that is accessible to rubber, since the internal micropore area is not accessible to rubber molecule. The results show that the average of BET surface area of r-CB is $44m^2/g$ while the STSA is $43m^2/g$. Compared with virgin carbon black, both results

Property	Minimum	Maximum	Test Result
Heating loss (moisture) (%) [7]	0.0	2.0	0.62
Ash content (%) [8]	5.0	20.0	15.8
Carbon Black Content (%) [9]	78.0	100.0	83.58
Comparison of surface area			
Type of carbon black	N330	N550	r-CB
BET SA (m^2/g) [10]	78	40	44.0
STSA (m^2/g) [10]	75	39	43.0

Table 1.
Certificate of Analysis (CoA) number: CA21-0137.

are higher than the surface area of N550, but still lower than N330. Based on these results, the r-CB is closer to N550, which is a semi-reinforcing grade of carbon black.

2.1 Rubber compound formulations

Table 2 shows the rubber compound formulations based on the Standard Test Methods for Styrene-Butadiene Rubber (SBR) Recipe and Evaluation Procedures [2]. Emulsion polymerised Styrene Butadiene Rubber (SBR 1502) was used. The gum compound was included so that one can assess and evaluate the magnitude of reinforcement attributed to r-CB and virgin carbon black. The "gum" or sometime known as "unfilled" refers to a rubber compound that does not contain any filler apart from the basic ingredients necessary for vulcanisation such as zinc oxide, stearic acid, sulphur and accelerator. All the units of the rubber compound formulations are expressed in parts per hundred of rubber (phr).

All rubber compounds shown in **Table 2** were prepared on a laboratory two-roll mill. Mixing was done in accord with the mixing procedures laid down in Standard Test Methods for Styrene-Butadiene Rubber (SBR) Recipe and Evaluation Procedures [2]. All test pieces were compression moulded at 145°C for 50 minutes.

2.2 Results and discussions

2.2.1 Hardness

Figure 1 shows the histogram of hardness for vulcanised rubber compounds tested according to the Standard Test Method For Rubbery Property-Durometer Hardness [12]. SBR is an amorphous rubber; therefore, it is an ideal rubber to evaluate the degree of reinforcement of filler. All the compounding ingredients and the quantity used were the same apart from different grades of virgin carbon black. Thus, the physical properties of the vulcanizate are attributed to the influence of the carbon black solely. Gum (unfilled) rubber acts as a reference of the strength associated with the cross-link network and viscoelastic effects. The quantity of carbon black used was fixed to 50 phr. Unfilled (gum) SBR has low hardness 41 Shore A. N330 gives 69 Shore A, an increase of 68.3% in hardness compared with gum vulcanizate. Overall the hardness of SBR compound filled with r-CB is lower than all the virgin carbon black filled SBR. Based on the results shown in

Reference	Gum (phr)	r-CB (phr)	HAF (N330) (phr)	FEF (N550) (phr)	GPF (N660) (phr)	SRF (N700) (phr)
SBR1502	100	100	100	100	100	100
ZnO	3	3	3	3	3	3
Stearic acid	1.0	1.0	1.0	1.0	1.0	1.0
r-CB	—	50	50	50	50	50
TBBS	1.0	1.0	1.0	1.0	1.0	1.0
Sulphur	1.75	1.75	1.75	1.75	1.75	1.75

Table 2.
Rubber compound formulations.

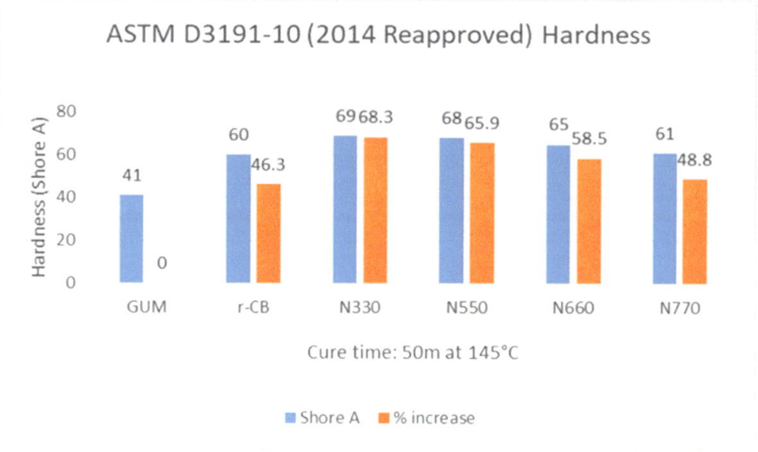

Figure 1.
A histogram showing the hardness of r-CB in comparison with virgin carbon black.

Figure 1, the degree of reinforcement of r-CB is close and comparable to N770. The reason for the low hardness is not entirely clear, but might be attributed to poor rubber-filler interaction because of the high level of ash content covering the r-CB particles. The ash content of r-CB is 15.8% as shown in **Table 1**; in contrast, virgin carbon black has very low ash content less than 1% [13]. However, r-CB still meets the hardness of 58 Shore A for pyrolysed carbon black for rubber as reported by William *et al* [14].

2.2.2 Tensile properties

The tensile test was done in accordance to the International Organisation for Standardisation (ISO) for Rubber, vulcanised or thermoplastic [15]. **Figure 2** shows the results of tensile stress at 100% strain (M100) and tensile stress at 300% strain (M300). The M100 and M300 of r-CB are still lower than those of all grades of virgin carbon black. The lower hardness, M100 and M300 of r-CB than those of virgin carbon black reflect the poor rubber-filler interaction. The contaminants (ash, oil, etc.) on the surface of r-CB might interfere in the surface activity of r-CB and prevent the formation of efficient linkages or interaction between the filler and the molecular rubber chains.

Figure 3 shows the tensile strength of r-CB against the virgin carbon black. The tensile strength of vulcanised SBR filled with r-CB meets the minimum requirement of 18.2 MPa for a black filler for rubber from tyre pyrolysis [14]. The tensile strength r-CB is 11 times higher than that of vulcanised unfilled (gum) SBR indicating a substantial reinforcement produced by r-CB. However, the tensile strength of r-CB is still slightly lower than that of N330 and N550. The tensile strength of SBR filled with N330 is 21% higher than that of r-CB and SBR filled with N550 is 13% higher than r-CB. The tensile strength of SBR filled with N770 is 4% lower than that of r-CB. The level of reinforcement of r-CB based on tensile strength is equivalent to semi-reinforcing carbon black N660. Vulcanised SBR filled with r-CB gives the highest elongation at break than all grades of virgin carbon black as shown in **Figure 4**. The low hardness, low M100 and M300 all contribute to high elongation at break of r-CB. The minimum requirement for elongation at break for pyrolysed carbon black is

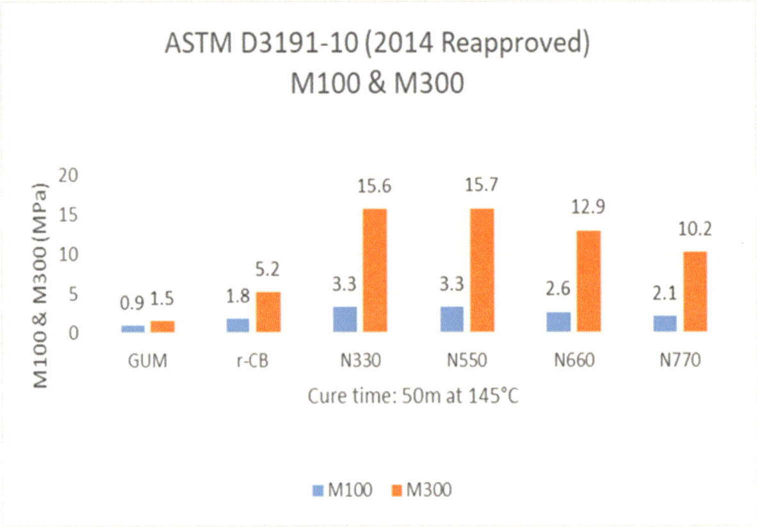

Figure 2.
A histogram showing the results of M100 and M300.

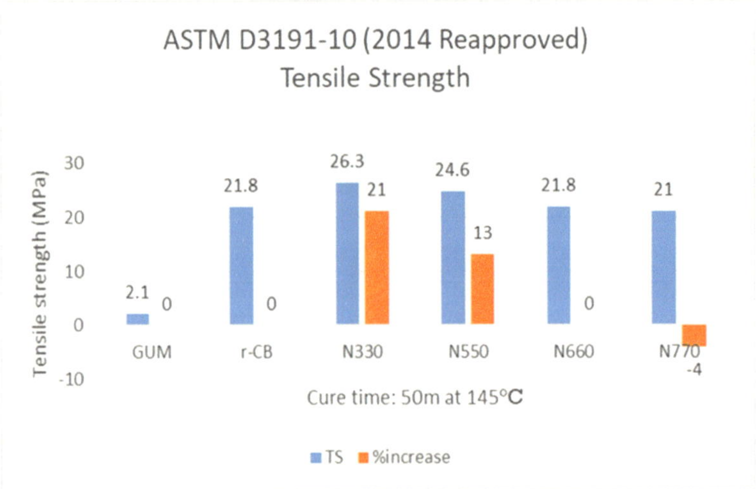

Figure 3.
A histogram showing the tensile strength of r-CB against virgin carbon black.

620% [14]. Again the r-CB used here meets the ASTM D3191 requirement for pyrolysed carbon black. The r-CB provides high resistance to deformation at high strain.

2.2.3 Tear strength

The tear strength was done by using trouser test pieces in accord with the ISO 34-1: 201 [16]. The results are shown in **Figure 5**. Although the tensile strength of r-CB is lower than N330 and N550, r-CB filled SBR produces the highest tear strength compared with all grades of virgin carbon black. Tear strength of r-CB is 520% higher than unfilled that of (gum) vulcanised SBR and 35.5% higher than that of black-filled

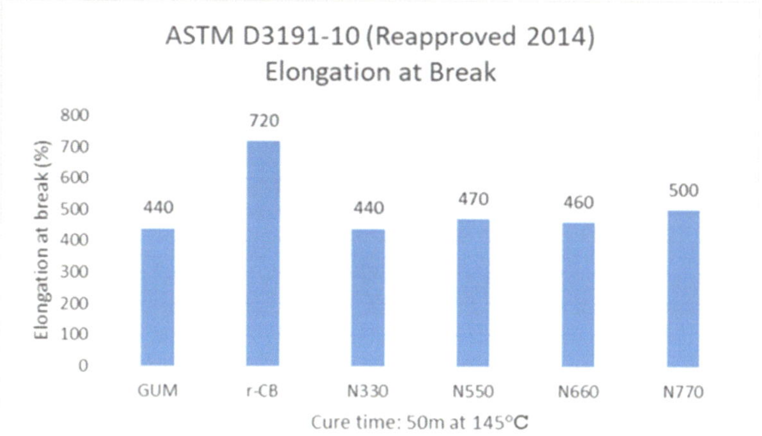

Figure 4.
A histogram showing the elongation at break of r-CB against virgin carbon black.

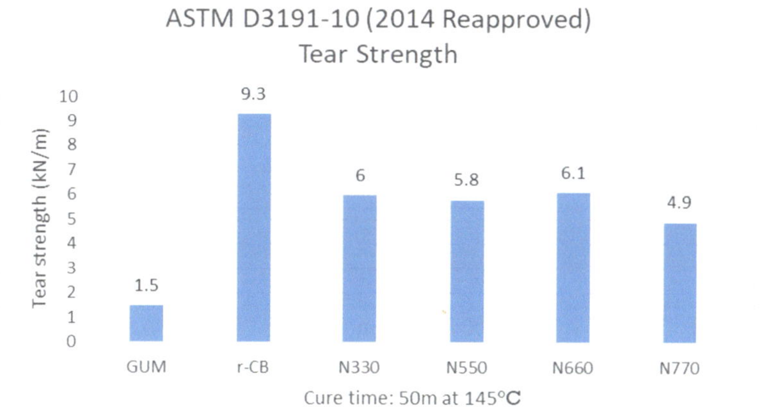

Figure 5.
A histogram showing the tear strength (trouser test-piece) of r-CB against virgin carbon black.

(N330) SBR. The tear strength does not correlate with the tensile strength because the failure mechanisms between the two are different. Tear failure involves crack growth process and greatly influenced by the energy dissipation process (hysteresis); in contrast, tensile failure is a single fatigue process.

The possible explanation why r-CB shows high tensile strength and high tear strength in spite of the low hardness, M100 and M300 is as follow: **Figure 6** shows a schematic diagram of r-CB. A virgin carbon black has very low ash content (less than 1%), and its surface activity is greatly influenced by the oxygenated functional groups such as carboxyls, phenols and ketones that are responsible for chemical interactions between the carbon black particles and the rubber molecular chains. In contrast, r-CB has high ash content (15.8% as shown in **Table 1**). The presence of ash and impurities on the surface of r-CB would interfere in the extent of wettability and the strength of the rubber-filler interaction. The ash content has been considered as a major barrier for r-CB to form a strong interaction with the rubber because the ash covers r-CB

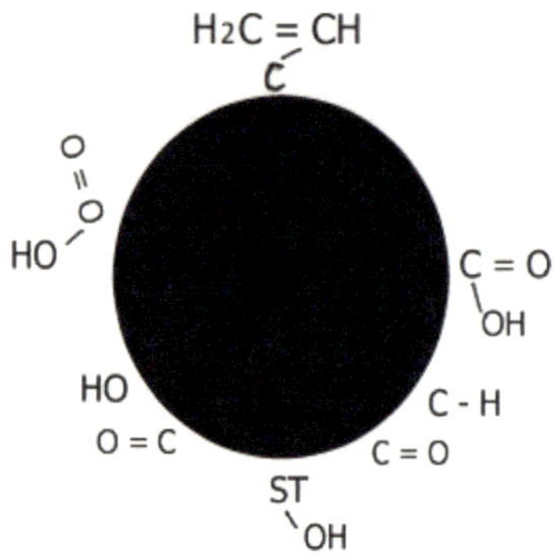

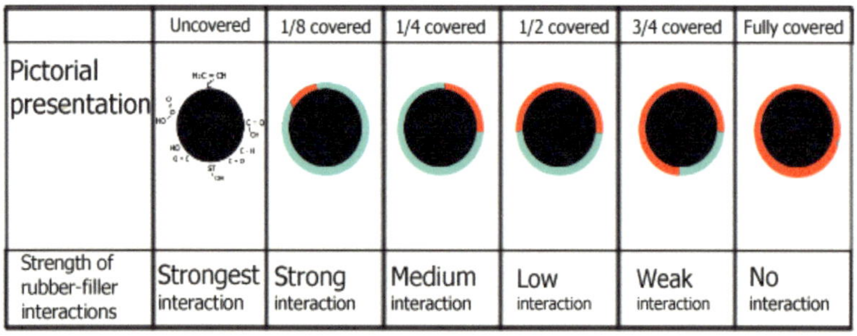

Figure 6.
Schematic diagram of r-CB showing the extent of impurities covering the r-CB particles. The red colour represents the impurities covering the r-CB particle. The red colour indicates the ash and impurities covering the carbon black particle, the light blue indicates the available space for rubber-filler interactions.

particles and hinders the interaction between r-CB and the rubber. The extent of the interaction is influenced by the availability of the area not covered by ash. Taking into two extreme cases, if the particle of r-CB is fully covered by ash, then there will be no interaction between the filler and the rubber. If the whole surface of r-CB is not covered by ash, then we have full interaction between filler and rubber. In reality, the amount of ash covering the surface of r-CB varies randomly. In the case of r-CB, the rubber-filler interaction might not be uniform, but a mixture weak and strong interactions.

There will be varying distribution of the strength of the interaction between filler and rubber depending on how much area is covered by ash and how much area is available for rubber filler interaction. In the regions where the ashes covering the r-CB particles are high, the rubber-filler interaction is low. It requires low stress to deform at a given strain level. For these reasons, the hardness, M100 and M300 are low in r-CB compared with virgin carbon black.

Tensile strength, elongation at break and tear strength involve high strain level reaching the breaking or fracture point. When the stress is applied, the weak interaction is broken first and then transfers the stress to next strong interaction. In so doing, the stress is relieved before reaching the breaking point, and the network is able to support further higher load. The process involves stress-relieving mechanism. The cycles repeat themselves until complete rupture occurs. The load transferring process also gives rise to energy dissipation or hysteresis. This is one of the reasons why r-CB provides higher tear strength compared with virgin carbon black.

2.2.4 Comparison of tensile strength and elongation at break at the same hardness level (60 Shore A)

Figure 7 shows the comparison of r-CB against Sterling V (Carbon black produced by Cabot), N550, N660 and N772 based on Cabot's Product Data Sheet [17, 18]. Cabot's compounding formulations were based on the Standard Test Methods for Styrene-Butadiene Rubber (SBR) Recipe and Evaluation Procedures [2], but the carbon black loading was adjusted to produce 60 Shore A hardness. The tensile strength

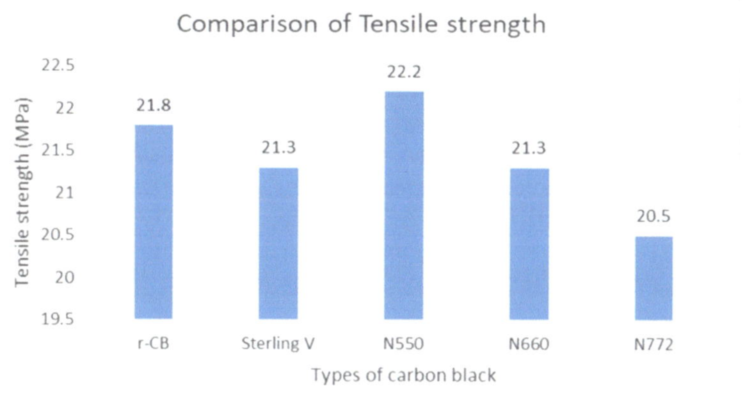

Figure 7.
A histogram showing the comparison of tensile strength of r-CB with Cabot's carbon black reported in the Technical Data. Comparison at the same hardness 60 Shore A.

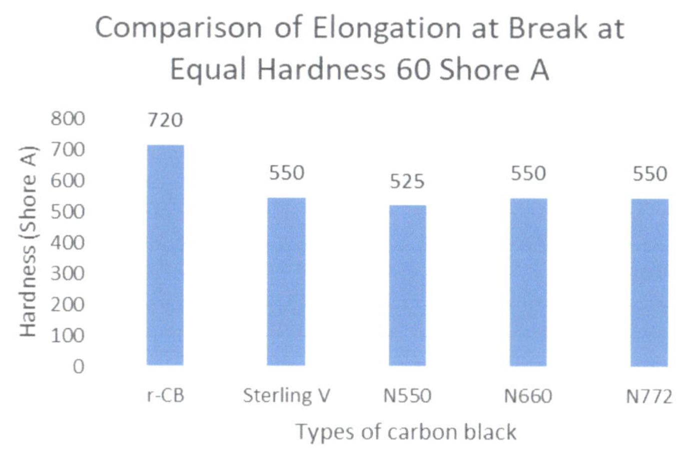

Figure 8.
A histogram showing the comparison of elongation at break at equal hardness 60 shore.

of r-CB black filled SBR is almost similar to N550 but slightly higher than Sterling V and N660 and that of N772.

Figure 8 shows the comparison of elongation at break at equal 60 Shore A hardness. It is clear that r-CB gives higher elongation at break by 31% compared with all the Cabot's virgin carbon black.

3. Evaluation of r-CB based on black-filled natural rubber (NR) standard formulation

Table 3 shows the basic black-filled Natural Rubber (NR) compound formulation. The Control compound was reproduced from the author's own work [19]. The aim here is to make relative comparison on the tensile properties of r-CB with that of virgin carbon black HAF (N330). The amount of r-CB used was 60 phr and would be compared against control compound filled with 50phr of HAF (N330) without processing additives. Compound AZBF1 contained processing additives such as Rubber Process Oil (RPO) and fatty acid zinc soap (FAZS) to facilitate filler dispersion and wettability. Rubber Process Oil was incorporated to facilitate flow and ease of mixing. The accelerator used was based on sulphonamides the fast curing, delayed action accelerators. The sulphur level was increased from 1.2 phr in the control compound to 1.3 phr for compounds AZBF1 and AZBF2. In compound AZBF2, the amount of r-CB was reduced to 50% to 30 phr with the addition of 3.5 phr of graphene Nanoplatelets. The aim here is to see whether graphene Nanoplatelets could enhance the tensile

Ref. no →	*Control (phr)	AZBF 1 (phr)	AZBF 2 (phr)
Natural Rubber (SMR 10)	100	100	100
HAF (N330)	50	—	—
r-CB	—	60	30
Graphene oxide	—	—	3.5
Fatty Acid Zinc Soap	—	1	—
Amide Fatty Acid Soap	—	1	—
Rubber Process Oil (RPO)	—	5	—
Zinc Oxide	5	5	5
Stearic Acid	2	2	2
N,N′-diaryl-p-phenylenediamines	—	1	1
Polymerised 2,2,4-trimethyl-1,2-dihydroquinoline	2	—	—
Morpholinylbenzonthiazole–2-sulphenamide	1.2	—	—
N-cyclohexylbenzothiazole-2-sulphenamide	—	1.2	1.2
Sulphur	1.2	1.3	1.3

**Reproduced from Reference [19]: A. B. Samsuri, Tear Strength of Filled Rubbers, PhD Thesis, Council for National Academic Awards, England, May 1989.*

Table 3.
Evaluation of tensile properties of r-CB based on standard NR black-filled compound.

Ref. no →	Control (phr)	AZBF 1 (phr)	AZBF 2 (phr)
Cure time @150°C (minutes)	25	8	8
Tensile Properties (ISO 37: 2017)			
M100 (MPa)	0.8	1.2	1.4
M300 (MPa)	1.64	4.7	4.1
Tensile strength (MPa)	26.9	24.7	27.2
Elongation at break (MPa)	749	750	780

Table 4.
Tensile properties of vulcanised r-CB filled NR.

properties of r-CB since it is well established that Nanofiller enhances the strength properties of polymer substantially. The Nanofiller (xGnP® Graphene Nanoplatelets Grade M) used was purchased from XG Sciences [20]. Grade M particles have an average thickness of approximately 6–8 nanometres and a typical surface area of 120–150 m^2/g. The average particle diameters of 15 microns. Preparation of the rubber compounds was done on a two-roll mill using the standard or normal mixing practice and procedures.

3.1 Results and discussions

The tensile properties are shown in **Table 4**. The NR filled with 50 phr of N330 gave tensile strength of 26.9 MPa about 8.9% higher than 60phr r-CB filled NR. In terms of elongation at break r-CB filled NR gives higher EB than 50 phr N330 NR filled although the former is filled with 60 phr r-CB, and the latter is filled with 50 phr N330. The reason might be associated with the presence of process in the 60 phr r-CB compound. When the r-CB loading was reduced to 30 and 3.5 phr of Nanoplatelet was added into AZBF2 compound, the tensile strength increases to 27.2 MPa. The tensile strength of r-CB filled with small quantity of graphene platelet is equivalent to the reinforcement produced by N330 HAF black although the quantity of r-CB is only at 30 phr. The degree of reinforcement is affected by the particle size, the filler structure, physical nature of the surface and the degree of filler dispersion [21]. The particle size of graphite is 0.34 nm; in contrast, the primary particle of carbon black is between 20 and 50 nm [22]. Particle size of graphite is 100 times smaller than that of N330 particle size. The small-size particle of graphite provides higher surface area than that of the particle size of N330 black. The high surface area enhances the ability of the filler to wet the rubber and enhances the interaction at the rubber-filler interface, thus giving very high reinforcement. The quantity of graphene required is only small because of the high surface area of the particle size.

4. Evaluation of r-CB in pre-cured tread compound

Table 5 shows the r-CB Natural Rubber (NR) masterbatch for the evaluation of pre-cured tread compound. Processing additives were used to facilitate mixing and ease fo filler dispersion. The total carbon black in each masterbatch (MB) was 60 phr. In the case of MB1, r-CB was mixed with virgin carbon black N220 (ISAF) in the ratio 50:50 (r-CB:N220). The aim was to see whether the addition of ISAF would enhance

Reference	MB1	MB2	MB3
Materials	phr	phr	phr
SMR10	100	100	100
r-CB	30	60	60
ISAF N220	30	—	—
Fatty Acid Zinc Soap	1	1	1
Amide Fatty Acid Soap	1	1	1
Rubber Process Oil (RPO)	5	5	—
Epoxidised Palm Oil (EPO)	—	—	5
Total	167	167	167

The abbreviations MB1, MB2 and MB3 refer to rubber compound masterbatch with different compounding ingredients.

Table 5.
r-CB NR masterbatch.

the tensile strength and hardness for the r-CB. In the case of MB2 and MB3, r-CB was used alone at 60phr loading. MB2 used RPO as the lubricant, and MB3 used EPO as the lubricant. EPO is more environmentally friendly than EPO since the former used waste palm oil instead of crude oil as the starting material. The full formulations for the pre-cured tread compound are shown in **Table 6**. Rubber compounds AZTC1–AZTC3 were based on 140 phr of rubber (NR 100 + BR 40phr). In 167 phr of MB, the NR content is 100 phr and the carbon black content is 60 phr. When 40 phr of BR is added, the carbon black content is diluted to (60/140 = 42.8phr. It was necessary to top up 24phr of carbon black so that the total carbon black is 60 parts. (60 + 24)/140 = 60 phr. For this reason, additional 24 phr of N220 was added in rubber compounds AZTC1 and AZTC3. In the case of AZTC2 rubber compound, 24phr of r-CB was added to produce total loading of 60phr.

For compounds AZTC4 and AZTC5, the MB1 and MB2 used were 102phr. In this case, the NR content was 60 phr and the carbon black content was 36 phr. It was necessary to top up 24 phr of either N220 or r-CB to make it to 60 phr when 40 phr of BR was added. The control compound was reproduced from the truck tyre retread formulation based on the Natural Rubber Formulary and Property Index [23] for the purpose of comparison. The control compound incorporated 70 phr of ISAF (N220) virgin carbon black.

The masterbatches shown in **Table 5** were prepared by using an internal mixer with batch weight of 2.5 kg. The fill factor used was 0.75. The internal has a fixed rotor speed of about 80 rpm. The NR was loaded first and premasticated for 1 minute before adding processing additives and half of the filler. Mixing was allowed to mix for another 1 minute before adding the remaining filler and oil. Mixing was allowed to continue until the temperature shown at the panel reached 120°C and the rubber masterbatch was discharged. The actual temperature of the masterbatch was measured by means of a thermocouple and recorded. The recorded temperature was in the range of 130–135°C. Rubber compounds shown in **Table 6** were prepared on a two-roll mill. All compounds shown in **Table 6** were cured at 150°C to t90, except the control compound where it was cured at 140°C for 20 minutes. The physical properties of pre-cured tread rubber compound produced by the local retreader are also included for comparison. The rubber compound of the retreader was made of blends

Sample reference	Control	AZTC 1	AZTC 2	AZTC3	AZTC4	AZTC5
OENR 75/25A	107	—	—	—	—	—
MB1 (r-CB 30: ISAF 30)	—	167	—	—	102	—
MB2 (r-CB 60)	—	—	167	—	—	102
MB3 (r-CB 60)	—	—	—	167	—	—
Polybutadiene rubber (BR)	20	40	40	40	40	40
r-CB	—	—	24	—	—	24
ISAF (N220)	70	24	—	24	24	—
Rubber Process Oil-2	—	—	—	—	5	5
Ppt. silica (VN3)	—	15	15	15	12	12
Coupling agent	—	1	1	1	1	1
Zinc oxide	5	5	5	5	5	5
Stearic acid	2	2	2	2	2	2
N,N′-diaryl-p-phenylenediamines	2	1.5	1.5	1.5	1.5	1.5
Wax blend	1	—	—	—	—	—
Polymerised 2,2,4-trimethyl-1,2-dihydroquinoline	—	1.5	1.5	1.5	1.5	1.5
Sulphur	1.5	1.8	1.8	1.8	1.6	1.8
N-t-butylbenzothiazole-2-sulphenamide	1.5	1.4	1.4	1.4	1.4	1.4
Diphenylguanidine	—	0.4	0.4	0.4	0.4	0.4

The abbreviations AZRTC 1–AZTC 5 refer to sample reference for five different tread compound formulations.

Table 6.
Pre-cured tread compound for tyre retreading formulations.

of NR, SBR and BR. The carbon black used was N234 (High structure ISAF) with 63 phr loading. Details of the formulations cannot be disclosed here [24].

4.1 Results and discussions

4.1.1 Hardness

The hardness results are shown in **Figure 9**. All compounds meet the Shore hardness specified by the Malaysian Standard Precured Tread for Retreading Tyres – Specification (3rd Revision): MS1208:2020 except AZTC4. Compound AZTC4 has higher N220 (42phr) than r-CB (18 phr) that exceeds the maximum hardness limit specified by the Malaysian Standard Precured Tread for Retreading Tyres – Specification.

4.1.2 M100 and M300

All compounds meet the M300 specified by the Malaysian Standard Precured Tread for Retreading Tyres – Specification the Malaysian Standard Precured Tread for Retreading Tyres – Specification as shown in **Figure 10**. The M300 exceeds the minimum value of 7.0 MPa. AZTC1 and AZTC4 produce high M300 because of the high

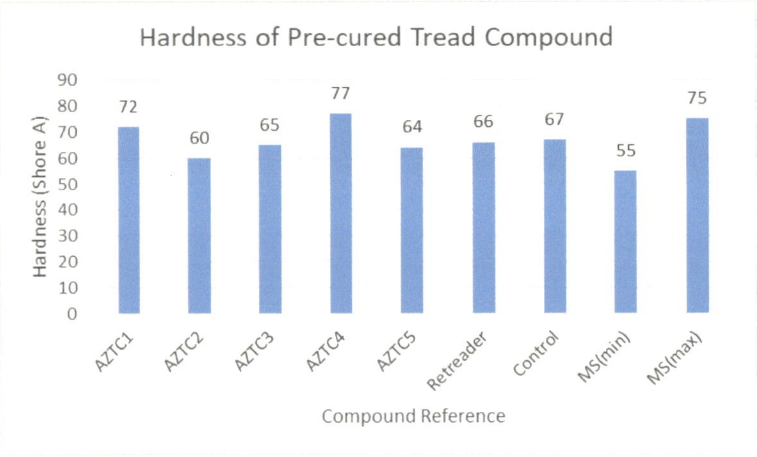

Figure 9.
Hardness of pre-cured tread compounds.

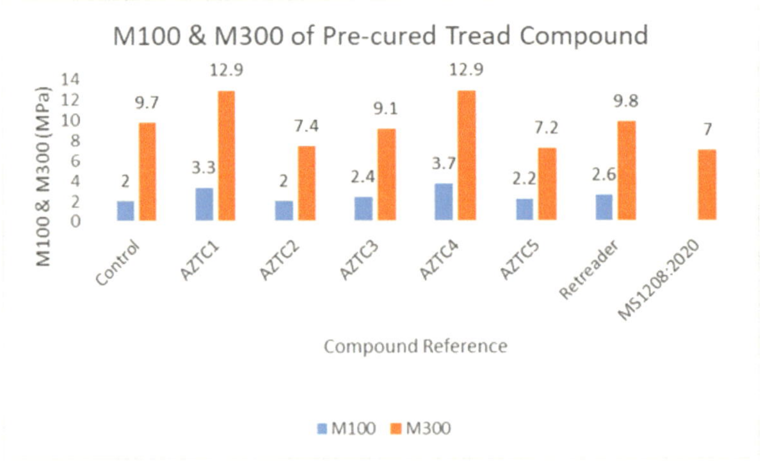

Figure 10.
M100 and M300 results of pre-cured tread compounds.

content of N220. However, the Malaysian Standard Precured Tread for Retreading Tyres does not specify M100.

4.1.3 Tensile strength

The tensile strength results are shown in **Figure 11**. All the compounds meet the minimum tensile strength (14.0 MPa) specified by the Malaysian Standard Precured Tread for Retreading Tyres. In fact, compounds AZTC1–AZTC4 produce higher tensile strength (12.2%–19.1%) than the retreader [24] that used 63phr of N234 and that of control compound that used 70 phr of ISAF (N220). The results show clear experimental evidence that r-CB can be used 100% on its own (AZTC2 and AZTC5) or blended with N220 (AZTC3) in pre-cured tread tyre compound and still meet all the requirements in the Malaysian Standard Precured Tread for Retreading Tyres.

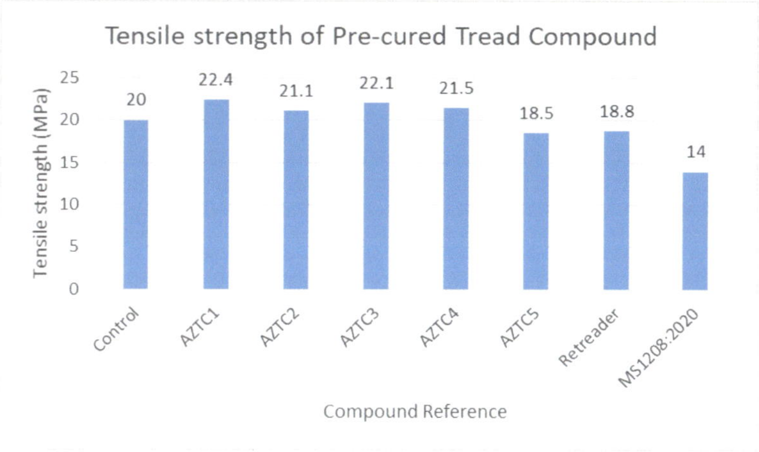

Figure 11.
Tensile strength of pre-cured tread compounds.

4.1.4 Elongation at break

The elongation at break results are shown in **Figure 12**. All the compounds meet the minimum elongation at break, EB (400%) specified by the Malaysian Standard Precured Tread for Retreading Tyres. Compounds having high content of N220 and low content of r-CB produce lower EB than compounds having more r-CB since the former gives higher hardness and M300 than the latter.

4.1.5 Ageing test results

The Malaysian Standard Precured Tread for Retreading Tyres specified ageing conditions at 70°C for 168 hours. The ageing test results for hardness, tensile strength and elongation at break are shown in **Table 7**. All the compounds meet the

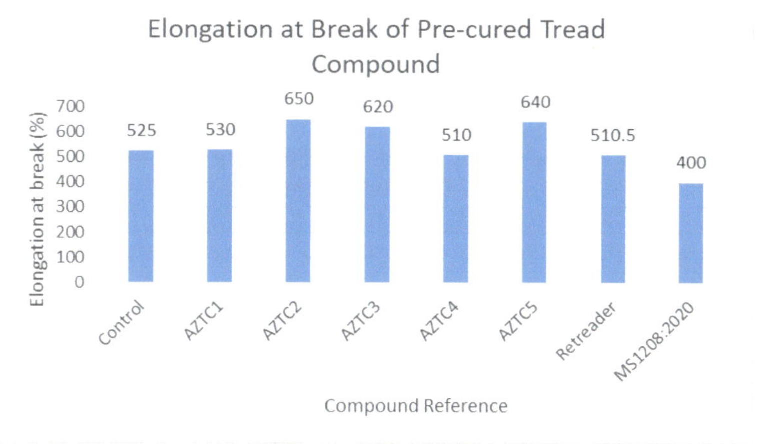

Figure 12.
Elongation at break of pre-cured tread compounds.

Sample reference	MS1208:2020	AZTC 1	AZTC 2	AZTC3	AZTC4	AZTC5
Hardness before ageing	55–75	72	60	65	77	64
Hardness after ageing	—	73	62	64	75	64
% change in hardness	+6	+1	+2	-1	-2	0
Tensile strength before ageing	14.0 MPa (min)	22.4	21.1	22.1	21.5	18.5
Tensile strength after ageing	—	18.8	18.6	20	20.5	17.1
% Retension in tensile strength	80	84	88	91	95	92
EB before ageing	400% (min)	530	650	620	510	640
EB after ageing	—	440	600	580	510	560
% Retension in EB	70	83	92	94	100	88

Table 7.
Ageing (70°C/168 h) test results for hardness, tensile strength and elongation at break.

MS1208:2020 specification after ageing at 70°C for 168 hours. The use of r-CB in the pre-cured tread compound does not affect the physical properties after ageing.

4.2 Abrasion resistance index (ARI)

The Abrasion Resistance Index (ARI) was done according to ISO 4649: 2017 (Method B) [25]. **Figure 13** shows the Abrasion Resistance Index (ARI). Abrasion resistance is a very important property in tyre because it determines the service life of the tyre in the long run. However, the Malaysian Standard Precured Tread for Retreading Tyres does not include this property in the specification. But we can compare with the ARI of the retreader. It is very surprising and interesting to note that the use of 100% r-CB (AZTC2 & AZTC5) gives comparable or even slightly better abrasion resistance than that of retreader that used 63phr of N234. Rubber compounds AZTC1 and AZTC4 give 30% and 47% higher than the retreader compound.

4.3 Tear strength

Trouser test pieces were used for the tear test [16]. **Figure 14** shows the tear strength results. Tear strength is another important property of a tyre.

It reflects the resistance to cutting by sharp and hard objects such as concrete curbs and sharp stones. All the results show comparable tear strength values.

4.4 Rebound resilience

Rebound resilience was done according to ISO 4662: 2017 [26]. **Figure 15** shows the rebound resilience results. Rebound resilience is another important property of a tyre. Resilience reflects the heat generation and rolling resistance of a tyre because it is influenced by the hysteresis or energy dissipation. High resilience indicates low

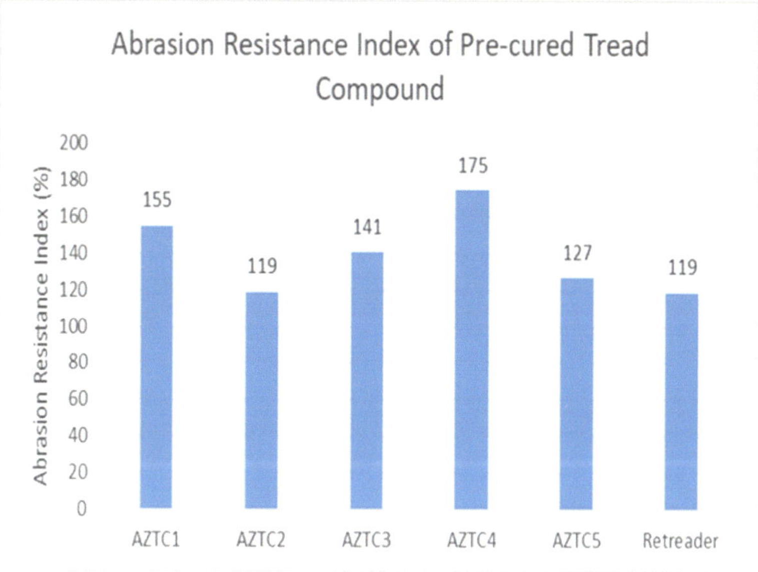

Figure 13.
Abrasion Resistance Index of pre-cured tread compounds.

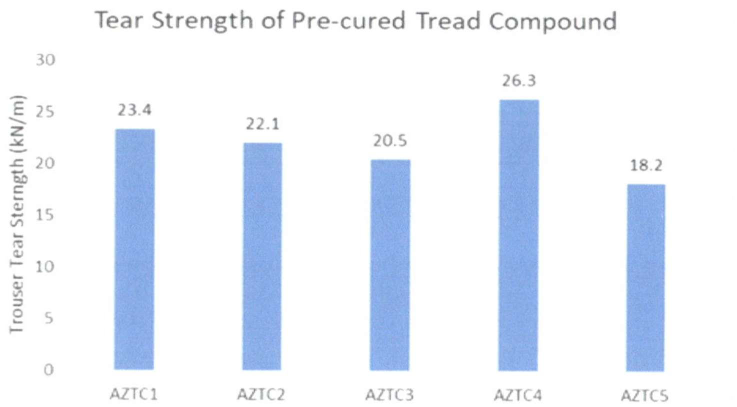

Figure 14.
Tear strength of pre-cured tread compounds.

rolling resistance and low heat build-up. All compounds having high r-CB produce higher resilience than compounds having high N220 and low r-CB. The retreader compound gives the lowest resilience because of the high loading of N234.

4.5 Summary

- The r-CB has high quality equivalent to virgin ISAF (N220).
- The incorporation of N220 into r-CB enhances further the mechanical strength and physical properties.

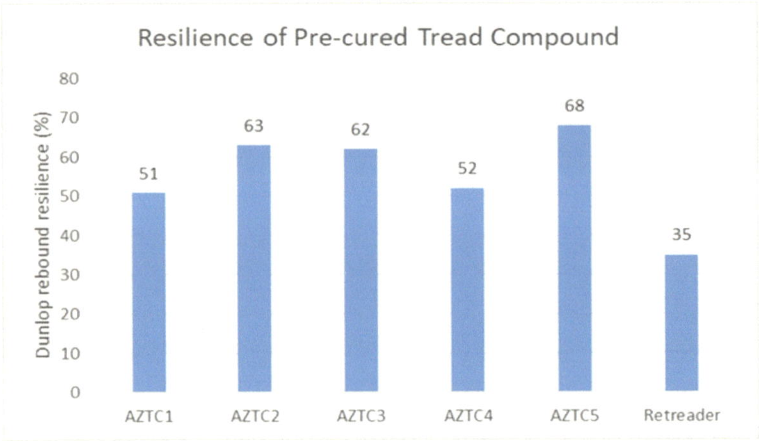

Figure 15.
Rebound resilience of procured tread compounds.

- r-CB can be used 100% on its own in pre-cured tread compound meeting all the specifications (both before and after ageing at 70°C 168 h) laid down in MS1208:2020.

- In addition, r-CB can provide cooling running tyre, low heat generation and low rolling resistance as reflected by the high rebound resilience. Low rolling resistance tyre saves fuel consumption and emits low smoke which are essential features of a green tyre that provides friendliness to the environment.

5. Applications of r-CB in automotive NBR compounds

Table 8 shows the rubber compound formulations based on Nitrile Rubber (NBR) for Rubber Seal such as O-ring and Inner Liner for Liquefied Petroleum Gas Lines. Rubber compounds AZOR1 and AZOR2 represent Rubber Seal (O-ring) formulations loaded with 100phr r-CB and 80phr r-CB respectively to replace 25phr of N-550 (FEF) and 95phr of N-990 (MT) in the Control 1 compound. Rubber compounds AZIL1 and AZIL2 represent Hose Inner Liner containing 100phr of r-CB and 80phr of r-CB respectively to replace 50phr of N-550 (FEF) and 50phr of N-990 (MT) in the Control 2 compound.

Rubber compounds shown in **Table 8** were prepared in a Farrel Banbury (BR1600) at a starting temperature of 70 °C, fill factor 0.7 and rotor speed of 100 rpm.

Curatives were added into the masterbatch the next day using a pre-heated two-roll at 50 °C. The mixing process took about 6 minutes.

5.1 Rubber seal (O-ring)

All the results are shown in **Table 9**. The ageing test was done at 100°C for 70 hours. The two NBR rubber compounds AZOR1 and AZOR2 meet the tensile strength and elongation at break requirements before ageing in accord with the ASTM Standard Classification System for Rubber Products in Automotive Applications [5]. However, rubber compound AZOR2 does not meet the minimum hardness

Materials/Ingredients	Control 1	AZOR1	AZOR2	Control 2	AZIL1	AZIL2
NBR (33% ACN)	100	100	100	100	100	100
N-550 (FEF black)	25	—	—	50	—	—
N-990 (MT black)	95	—	—	50	—	—
Carbon black (r-CB)	—	100	80	—	100	80
Stearic acid	0.5	0.5	0.5	1	1	1
Polymerised 2,2,4-trimethyl-1,2-dihydroquinoline	2	2	2	2	2	2
ZMBI	2	2	2	2	2	2
Dioctyl Phthalate	10	10	10	20	20	20
Zinc oxide	15	15	15	5	5	5
Processing aid Ultra flow 600 T	—	2	2	2	2	2
Sulfasan R	1.5	1.5	1.5	1	1	1
N-cyclohexylbenzothiazole-2-sulphenamide	3	3	3	3	3	3
Tetramethylthiuram disulphide	4	4	4	3	3	3
MC Sulphur	0.3	0.3	0.3	—	0.2	0.2

AZOR1–AZOR2 and AZIL1–AZIL2 refer to sample references for O-ring (OR) and inner liner (IL).

Table 8.
Nitrile rubber compounds for rubber sheet gasket and hose inner liner.

Sample reference	Control 1	AZOR1	AZOR2	Specifications [5]
Hardness before ageing (Shore A)	75	68	63	65–75
% change in hardness after ageing	6	7	5	0 (min) - +15 (max)
Tensile strength before ageing (MPa)	10	14.9	11.6	10.3 (min)
% Retension in tensile strength after ageing	+2	−16	−25	−15 (max)
EB before ageing	190	440	480	200 (min)
% Retension in EB after ageing	−34	−36	−41	−40 (max)
Compression set at 22 h/100 °C (%)	26	37	34	50 (max)
IRM 903 Oil immersion 70 h/100 °C	2	3	3	9 (max)

Table 9.
Physical properties before and after ageing for rubber seal.

requirement. It appears that 80 phr of r-CB is not adequate to meet the minimum hardness requirement of 65 Shore A. After heat ageing at 100°C for 70 hours, rubber compound AZOR1 meets the hardness and elongation at break requirements, but narrowly exceeds the maximum limit of the tensile strength retention. Both compounds AZOR1 and AZOR2 meet the compression set requirement below the maximum value of 50%. The ASTM Standard Classification System for Rubber Products in Automotive Applications specification for the change in volume after immersion in IRM903 oil for 70 h/100°C is +10% (maximum) [5]. IRM 903 oil (Formerly known as ASTM Oil No 3) is a testing oil to be used in testing rubber components as per ASTM standards. Both compounds AZOR1 and AZOR2 also meet the specification limit for

Sample reference	Control 1	AZIL1	AZIL2	Specifications [4]
Tensile strength before ageing (MPa)	10.8	13.3	10.3	9.81 (min)
Tensile strength after ageing (MPa)	10.9	11.5	8.8	
% Retension in tensile strength after ageing	100	86.5	85.4	75 (min)
EB before ageing (%)	370	670	660	200 (min)
EB after ageing (%)	292	563	548	
% Retension in EB after ageing	78.9	84	83	75 (min)
Swelling in n-pentane 72 h/20 °C	8	11	11	25 (max)

Table 10.
Physical properties of hose inner liner before and after ageing at 100°C for 70 hours.

swelling resistance in oil immersion where the change in volume after immersing in IRM903 oil for 70 h/100°C is only +3%.

5.2 Hose liner compounds

Table 10 shows the results of inner liner hose rubber compounds to meet the British Standard Specification for Rubber Hose and Hose Assemblies for Liquefied Petroleum Gas Lines [4] before and after ageing at 70°C for 96 hours. Both rubber compounds AZHL1 and AZHL2 meet the minimum requirements of tensile strength and elongation at break before and after ageing. In fact, the tensile strength of AZHL1 before ageing is 35.6% higher than the minimum tensile strength specified and its elongation at break is 235% higher than the minimum EB specified. The retention of tensile strength and retention of elongation at break after ageing are more than 80%. All the compounds also meet the change in volume after immersion in n-pentane for 72 hours at 20°C. The results above provide experimental evidence that r-CB reported here is of high quality and suitable for applications in the manufacturing of rubber products such as pre-cured tread liner, rubber seals (O-ring) and hose inner liner.

The prospects of r-CB are very bright to penetrate into the rubber industry market. With the escalating price of virgin carbon black, r-CB is the best alternative substitution of black filler because the cost of r-CB is around 40–50% cheaper than most of the grades of virgin carbon black. The challenges ahead are to maintain high quality of r-CB, to be competitive in cost in relation to virgin carbon black, to reverse the misconception of r-CB as a low-quality material with the view to promote and to gain acceptance of r-CB as a suitable quality material by the Rubber Industry players worldwide. The future of r-CB is very bright with the acceptance of big tyre manufacturers to use r-CB in their tyre rubber compounds.

6. Conclusions

- The recycled carbon black used in this investigation is of high quality approaching the properties of reinforcing carbon black N330 (HAF black).

- r-CB meets the tensile strength minimum requirement of 18.2 MPa for a black filler for rubber from tyre pyrolysis specified by the ASTM D3191.

- At equal hardness (60 Shore A) based on the Standard Test Methods for Styrene-Butadiene Rubber (SBR) Recipe and Evaluation Procedures, the tensile strength of r-CB is equivalent to N550 and higher than that of the commercial virgin carbon black such as Stirling V, N660 and N770.

- The r-CB can be used 100% on its own or in combination with N220 (ISAF) to produce pre-cured tread compound for retreading and meeting all the technical requirements specified in the Precured Tread for Retreading Tyres-Specification, The Malaysian Standard MS1208:2020.

- The r-CB can be used to replace virgin carbon blacks in NBR compounds to produce rubber seal such as O-ring and hose inner liner for rubber hose and hose assemblies for liquefied petroleum gas lines.

- The prospects and future of r-CB are very promising and bright with very competitive price about 40–50% cheaper than most of the commercial virgin carbon black.

- The biggest challenge is to produce high-quality r-CB with high assurance of consistency and uniformity.

Acknowledgements

The author would like to express his sincere gratitude and appreciation to Mr. Tojo Chang, the Managing Director of Eco Power Sdn Bhd, who has sponsored and supported the R&D work presented in this paper. Special appreciations also go to Dr. Ahmad Faiza Mohd and Dato' Yew for useful discussions and last but not least to Dr. Wong and Ms. Syazlin for carrying out the experimental work.

Author details

Azemi bin Samsuri
Department of Ranking, UiTM Global, Universiti Teknologi MARA, Shah Alam, Malaysia

*Address all correspondence to: azemi.sam@gmail.com

References

[1] Available from: https://hal-mines-albi.archives-ouvertes.fr/hal-01915193 July 13, 2021

[2] ASTM D-3191-10. Standard test methods for SBR (styrene-butadiene rubber)-recipe and evaluation procedures. (Reapproved 2014)

[3] Precured tread for retreading tyres-specification, The Malaysian Standard MS1208:2020 (3rd Revision)

[4] British Standard Specification for Rubber Hose and Hose Assemblies for Liquefied Petroleum Gas Lines, B.S. 4089. 1966

[5] ASTM Standard classification system for rubber products in automotive applications; designation D2000: 18

[6] Available from: https://www.bridgestoneamericas.com/en/newsroom/press-releases/2021/bridgestone-michelin-joint-presentation-on-rcb).

[7] ASTMD1509. Standard Test Methods for Carbon Black-Heating Loss. ASTM International (ASTM); 2018

[8] ASTM-D1506. Standard Test Methods for Carbon Black-Ash Content. ASTM International; 2015

[9] ASTMD3172. Standard Practice for Proximate Analysis of Coal and Coke. ASTM International; 2021

[10] ASTMD6556. Test Method for Carbon Black-Total and External Surface Area by Nitrogen Adsorption. ASTM International; 2021

[11] Walton KS, Snurr RQ. Applicability of the BET method for determining surface areas of microporous metal-organic frameworks. Journal of the American Chemical Society. 2007;**129**(27):8552-8556

[12] ASTMD2240-15e1. Standard Test Method For Rubber Property-Durometer Hardness. 2015

[13] Orion engineered carbons GmbH, Orion what is carbon black, general physical and chemical properties. 2015. p. 25

[14] Klingensmith W and Beck M. A black filler for rubber from tire pyrolysis char. American Chemical Society Meeting Washington DC, August 23, 1994. p. 30

[15] ISO 37:2017. Rubber, vulcanized or thermoplastic – Determination of tensile stress-strain properties. 2017

[16] ISO34-1:201. Rubber, vulcanized or thermoplastic – determination of tear strength – standard – standard test for trousers. Angle and Crescent Test Pieces. 2010

[17] Cabot product data sheet, sterling SO carbon black, 09/2014

[18] Cabot product data sheet, sterling V carbon black, 09/2014

[19] Samsuri AB. Tear strength of filled rubbers, [PhD thesis], council for National Academic Awards, England, May 1989

[20] Technical data sheet, 2013 XG sciences, Inc xGnP® grade M product characteristics

[21] Samsuri AB. Theory and mechanisms of filler reinforcement in natural rubber, natural rubber materials. In: Thomas S, Chan CH, Pothen LA,

Joy J, Maria HJ, editors. Composites and Nanocomposites. Vol. 2. RSC Publishing; 2014. Ch. 3. p. 73

[22] Vilgis TA. Reinforcement of Polymer Nano-Composites, Theory, Experiments and Applications. 2009. pp. 1-209

[23] The Natural Rubber Formulary and Property Index. The Malaysian rubber producers' research association. ISBN: 0-9504401-3-2. 1984

[24] Private Communications with GIIB RUBBER COMPOUND SDN BHD, Mac. 2022

[25] ISO 4649:2017. Rubber, vulcanized or thermoplastic (Mehod B), Determination of the resistance of rubber to abrasion by means of a rotating cylindrical drum. 2017

[26] ISO 4662:2017. Rubber, vulcanized or thermoplastic, Determination of rebound resilience. 2017

Chapter 3

Composition and Structure Influence on Properties of Elastomeric Composites with Silicon Dioxide Fillers

Yury V. Kornev, Hammat H. Valiev, Yuliya N. Karnet, Alla A. Kornilova, Nikolay A. Semenov and Alexander N. Vlasov

Abstract

The problem of finding effective polymer elastomer fillers now is actual task. This chapter presents experimental studies of composites based on industrial synthetic butadiene-styrene rubber with amorphous silicon dioxide filler particles, obtained from rice husk ash processing products. The new methodology was developed for producing highly dispersed fillers powders with micro- and nano-sizes using a planetary ball mill. The synthesized composites surface structure was investigated using scanning electron and atomic force microscopy. The composites elastic-strength properties were studied by tensile testing machine. The significant influence of the surface functionality and the carbon/silicon dioxide ratio in fillers on the elastomeric composites mechanical properties is shown. It is concluded that these new classes of reinforcing fillers can be recommended for practical use.

Keywords: elastomeric composites, amorphous silicon dioxide filler, rice husk ash, scanning electron and atomic force microscopy, mechanical properties

1. Introduction

Rubber materials belong to one of the main substances in modern industrial society [1, 2]. The main substitute of natural rubber synthetic styrene butadiene rubber (SBR) needs the proper fillers to attain the demanded performances [3]. Traditional reinforcement fillers such as black carbon and precipitated silica have many disadvantages. Black carbon produced from petroleum is toxic for environment, and as the precipitated silica have high cost in the production. Therefore searches to alternative fillers for SBR are the urgent task. Great interest aroused in the possibility to use as rubber fillers the multi-tonnage agricultural waste in rice production - rice husk ash (RHA), which contains amorphous silica. But many researchers conducted during decades in this direction yet were unable to create the SBR composites permissible for practical applications [4–24]. This Chapter summarizes our works in successful syntheses such

SBR composites based on RHA fillers and the conducted investigations demonstrate the promise possibilities for exploration these composites in rubber industry. Chapter includes the development in principles for creating composite materials based on polymer and elastomeric matrices and a new class of micro and nanostructured mineral fillers from rice husk ash (RHA) [25, 26], as well as the study of mechanical behavior features and reinforcement effects in such composites in order to directly regulate their properties. Based on experiments, it is proposed to use as a renewable raw material for the production of new types reinforcing fillers based on silicon dioxide – rice husk ash processing products (RHAPP). The main components of the obtained micro and submicron fillers are amorphous silica and carbon. This filler type differs in structure from the widely used carbon black and colloidal silica, but is similar to them in chemical composition and is obtained from renewable raw materials. In the course of obtaining highly dispersed fillers from RHAPP, the stage of ultrafine grinding is expected, which reduces energy costs and makes the technology more environmental safety in comparison with existing ones [27]. The technology was developed in our Institute and is used in the present work to obtain reinforcing fillers for investigated elastomeric composites. This Chapter presents a study of the filler particle size influence obtained from the RHAPP on the surface structure and mechanical properties of elastomeric composites based on a rubber copolymer butadiene - styrene SBR-30 ARK. In the proposed work, submicron particles of amorphous silicon dioxide were obtained for the first time by ultrafine grinding RHA. Also for the first time, the effect of reinforcing with mineral filler from RHAPP for elastomeric composites based on SBR-30 ARK was established. The relationship of the structure and complex of mechanical properties for elastomeric composites with a new class of strengthening mineral fillers has been established. These results are important not only for practical applications, but also for understanding the fundamental composition-structure–property relationships in rubber composites.

2. Experimental procedure and materials

The synthesized elastomeric composites were based on a copolymer butadiene - styrene SBR-30ARK with 30% styrene (Sibur, Russia, analogous SBR –1500). The rubber ingredients (in weight percentage) are shown in **Table 1**.

In **Table 1** rubber formulation presents various chemical additives to improve the vulcanization speed and efficiency and to control the cross-linking nature. Additives: Sulfur; Zinc oxide; Sulfenamide C: N-Cyclohexylbenzothiazol-2-sulfenamide (CBS); Altax: Mercaptobenzothiazole disulfide (MBTS); Stearin technical – are the products Bina Group LLC, Moscow, Russia. NORMAN 346 - special process oils with the lowest carcinogen content, polycyclic aromatic hydrocarbon (PAH) normalized filler oil for TDAE synthetic rubbers (the type Vivatec 500) - the product AO GC Titan, Omsk, Russia.

Amorphous silicon dioxide microdispersed and submicron (nano) particles obtained from rice husk ash processing products (RHAPP) were used as the filler.

Substance	SBR-30ARK	Sulfur Sulfenamide C Altax	Zinc oxide	Norman 346	Stearin technical
%	100	4.7	3	7	2

Table 1.
The butadiene - styrene SBR-30ARK rubber ingredients.

It includes white amorphous silicon dioxide samples (up to 97%) and carbon impurities (3–5%) and one black RHAPP sample with 35% amorphous silicon dioxide and 60% graphite-like carbon (LTD Leader 2004, Russia). To increase the specific surface area of the rice husk ash and, consequently, its reinforcing activity in elastomeric composites, the initial fraction was subjected to additional grinding by a special method developed at the Institute of Applied Mechanics, Russian Academy Sciences (IAM RAS) (RU patent No. 2442657). The temperature of the initial mixture does not exceed 150°C. Submicron particles were also used in composites after oxidative surface treatment with hydrogen peroxide H_2O_2. For better filler particles dispersion in the elastomeric matrix volume was additionally used the coupling agent (1.5 mass % surface modifier) organosilane (bis (3-triethoxysilylpropyl) tetrasulfide) - TESPT - Sigma-Aldrich (Merck) as in [28]. The double modification – combination oxidized H_2O_2 submicron particles with TESPT was also used. The prepared and studied samples components of elastomeric composites are shown in **Table 2**.

In **Table 2** symbols correspond to the following composites: sample № 1: pure SBR-30ARK without fillers; sample № 2: filler RHAPP micro US— RHAPP micro particles, ultrasound treatment; sample № 3: filler RHAPP nano — RHAPP ultrafine grinding (sub micro- nano particles); sample № 4: filler RHAPP nano H_2O_2— as № 3, additionally processed by oxidation with H_2O_2; sample № 5: filler RHAPP nano TESPT – as № 3, additionally processed by TESPT; sample № 6: filler RHAPP nano H_2O_2/TESPT – as № 4 additionally processed by TESPT; sample № 7: filler RHAPP black micro US —RHAPP micro particles, ultrasound treatment.

For results comparison convenience, the mass concentration for all fillers was taken 65 phr (filler parts per hundred rubber parts in weight). The fillers dispersion and blending all elastomers ingredients was carried in a laboratory mixer Haake PolyLab 300 QC (Thermo Scientific, Dreieich, Germany). The mixing temperature did not exceed 150°C. Control of the particle size and morphology before and after grinding was performed using field emission scanning electron microscopy (SEM) JSM-6700F (JEOL, Tokyo, Japan) with a JED-2300F energy dispersive microanalyzer attachment. Tests were conducted in high vacuum mode. Preliminary platinum sputtering no thicker than 50 nm was applied on the surface of samples. The accelerating voltage during the experiments did not exceed 20 kV. Also was used transmission electron microscopy (TEM) (FEI Tecnai G2 F20 FEG). The atomic-force microscope (AFM) easyScan DFM (Nanosurf, Switzerland) operating in the semi-contact mode under normal conditions with additional phase contrast mode and SuperSharpSilicon probe (Nanosensors, Switzerland) with a tip radius of about 2 nm were used. Additional

The composite samples filler ingredients	№ 1	№ 2	№ 3	№ 4	№ 5	№ 6	№ 7
Pure SBR-30ARK	100	100	100	100	100	100	100
RHAPP micro US	0	65	0	0	0	0	0
RHAPP nano	0	0	65	0	0	0	0
RHAPP nano H_2O_2	0	0	0	65	0	0	0
RHAPP nano TESPT	0	0	0	0	65	0	0
RHAPP nano H_2O_2/TESPT	0	0	0	0	0	65	0
RHAPP black micro US	0	0	0	0	0	0	65

Table 2.
The prepared and studied elastomeric composites samples components.

ACM protection from external excitations was provided by means a dynamic anti-vibration stage TS-150 (Fabrik am Weiher, Switzerland). The surfaces image processing was performed using the SPIP™ - advanced software package for processing and analyzing microscopy images at nano- and micro scale (Image Metrology, Denmark). The elastomeric composites mechanical characteristics were determined in accordance with ASTM [29] on the UTS-10 (Zwick Roell, Germany) universal testing machine. The tests were carried out at ambient temperature with a constant tension rate of 100 ± 10 mm/min. During the experiment, the engineering stress σ (MPa) – strain ε (%) curves were obtained. The complex effective component dynamic shear modulus G'(Pa) dependence on the relative strain for rubber compounds was carried out on a device HAAKE RheoStress RS 150 (Haake, Germany) at a frequency 10 Hz and temperature 60°C, in the deformation amplitude range 0.001–0.3.

3. Results and discussion

The particle size and morphology before RHA and after grinding RHAPP was performed using SEM (**Figure 1a,b**). It is noted that the initial particle size RHA was in range 10–300 microns and have a porous layered structure (**Figure 1a**). After grinding there is a significant decrease in the RHAPP (nano) particle size (**Figure 1b** compared to **Figure 1a**); and it is established that the nanosized particles are capable for strong agglomeration. It was found that the average particle size of the obtained highly dispersed fillers is up to 700 nm. The individual particles TEM images of rice husk ash processing products after grinding (RHAPP nano) are presented in **Figure 2**.

According TEM results the individual sizes RHAPP particles are in the range 10–20 nanometers. The surface scans of elastomeric composite sample № 3 (filler RHAPP nano) obtained during the SEM experiments are shown in **Figure 3**. The image visualize filler particles inclusions (lighter areas) in the elastomeric matrix volume (darker areas). The inclusion sizes in the samples filled with these sub-microfine fillers were much smaller and indicate markedly better filler particles dispersion in composite. Thus according the SPIP analysis for composite № 3 mean inclusion sizes are in the range 30 nm and mean distances between inclusions about 100 nm, while for example in composite № 2 (filler RHAPP micro) these are 200 nm and 450 nm accordingly.

In **Figures 4–9** examples AFM surface images on samples № № 2–7 from the synthesized composites listed at **Table 2** is displayed. In topography scans the filler particles are visualized as light areas in the elastomeric matrix dark areas. In **Figure 4** the filler agglomerate sizes are in the range 300 nm – 3 μm. Phase contrast image shows rather homogeneous microfiller distribution in rubber matrix with formation of certain filler structuring in matrix. It is visualized that the boundary layers of polymers adsorbed on the filler surface overlap. This testifies the filler ultrasound treatment success before its introduction into the elastomer, which provided the filler homogeneous size distribution in rubber. Thus, we can assume that the first condition for the emergence of the rubber reinforcement effect by the filler is fulfilled.

However, as will be shown by elastic strength characteristics studies of this composite, this condition is not sufficient, definitely due to the fact that the filler particles are micro-sized; this is the reason for poor adhesion of the filler to the matrix, and consequently, for the composite's physical and mechanical properties.

In **Figure 5** the filler agglomerate sizes are in the range 50 nm – 100 nm. The phase contrast image shows absence the filler coherent cooperative structuring: filler separately, matrix separately. Obviously, it is because the initial silica, even in the

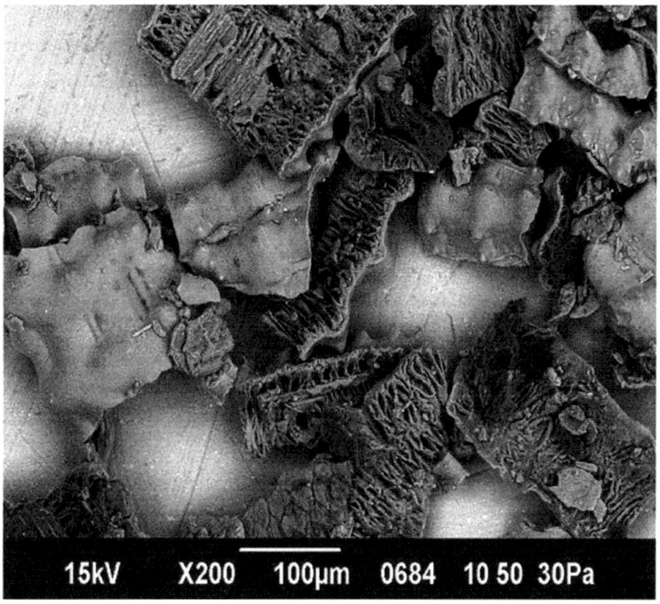

(a)

(b)

Figure 1.
SEM images rice husk ash particles before and after ultrafine grinding: a) RHA (micro) initial particles, b) RHAPP milled (nano) particles agglomerates.

nanoscale range, is hydrophilic whereas the rubber matrix is hydrophobic and hence there is very little interaction between them.

In **Figure 6** the filler agglomerate sizes are in the range 100 nm – 3 μm. According to phase contrast image the cohesion, cooperation and structuring of filler is

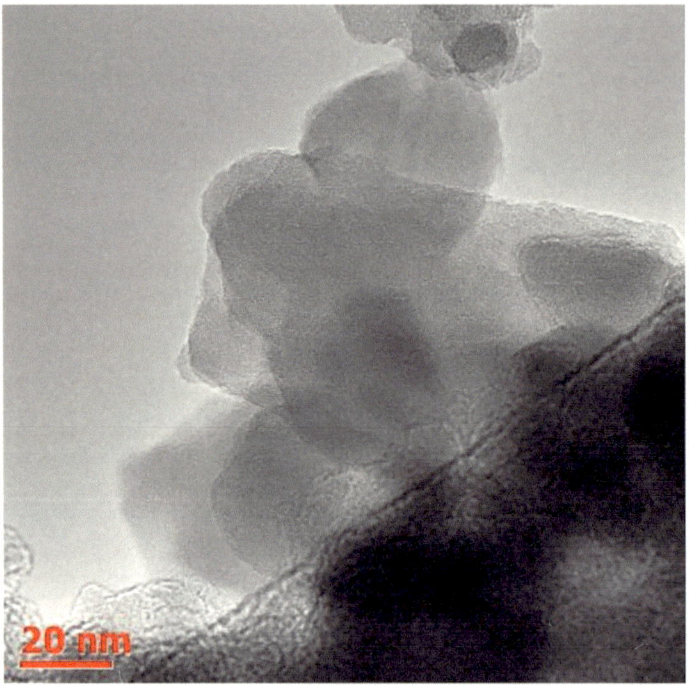

Figure 2.
The individual particles TEM images of rice husk ash processing products after grinding (RHAPP nano).

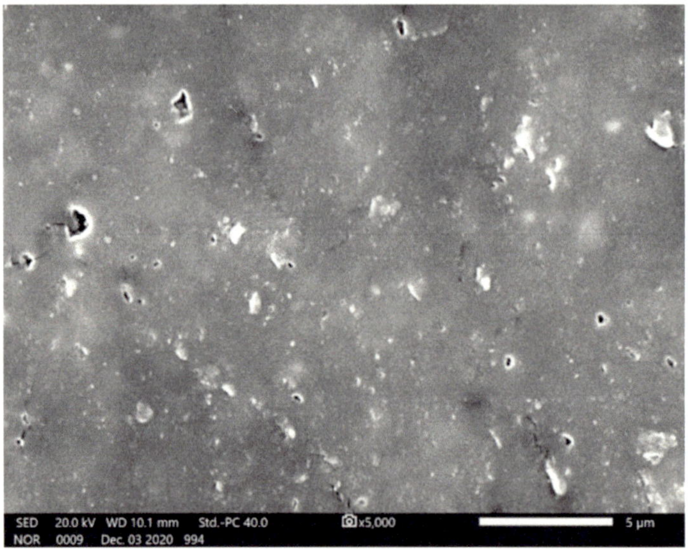

Figure 3.
The SEM surface morphology image of elastomeric composite with submicrodispersed amorphous silicon dioxide fillers. Sample № 3 (filler RHAPP nano).

observed; however, here large filler agglomerates down to micron size were formed, despite that the initial filler size introduced into matrix was in the nano range. This indicate that the silica-to-silica interaction is strong due to hydrogen bonding by

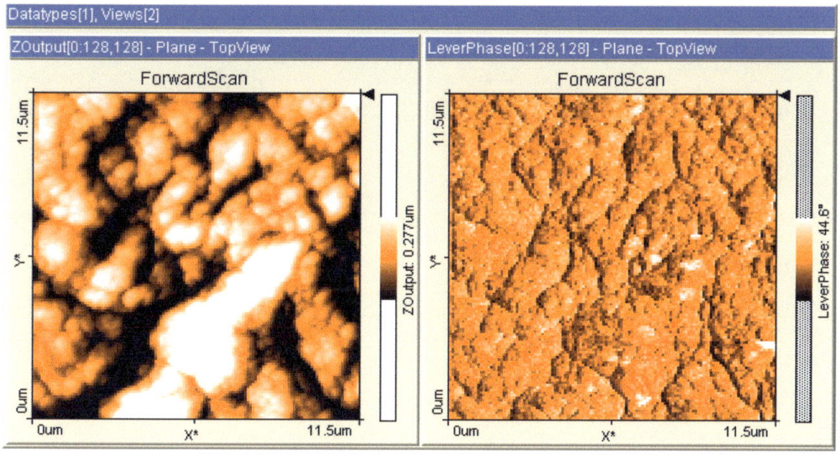

Figure 4.
AFM surface images composite № 2 (RHAPP micro). Scans 11.5 × 11.5 μm². Left - topography, right - phase contrast.

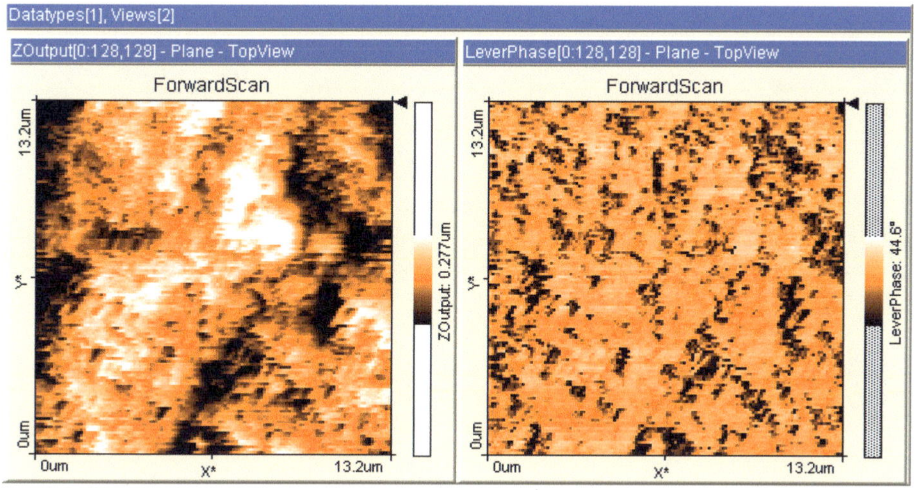

Figure 5.
AFM surface images composite № 3 (RHAPP nano). Scans 13.2 × 13.2 μm². Left - topography, right - phase contrast.

silanol groups on silica surfaces, which results in relatively more aggregation of silica particles in the rubber matrix.

In **Figure 7** the filler agglomerate sizes are in the range 100–700 nm. Phase contrast image showed good dispersion and distribution of nanofiller in matrix in a cooperative, coherent and structured manner. Obviously, this is due to the use of TESPT.

In **Figure 8** the filler agglomerate sizes are in the range 100–700 nm. Analogous to composite No. 5 pattern is visualized in this composite No. 6 phase image. Thus, in these composites No. 5 and No. 6 a significant increase in mechanical properties would be expected. All these conclusions were confirmed by direct examination of the mechanical properties of the composites on a tensile testing machine.

In **Figure 9** the filler agglomerate sizes are in the range 300 nm – 2.5 μm.

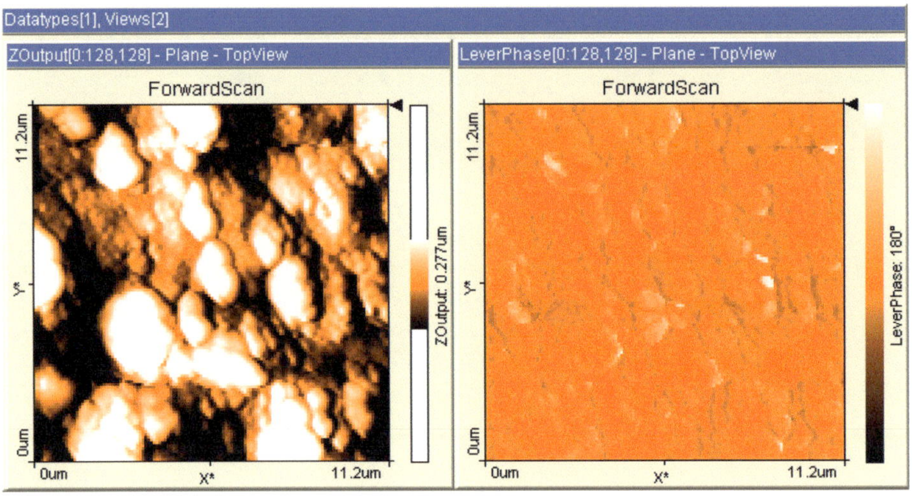

Figure 6.
AFM surface images composite № 4. (RHAPP nano, oxidized). Scans 11.2 x 11.2 μm². Left - topography, right - phase contrast.

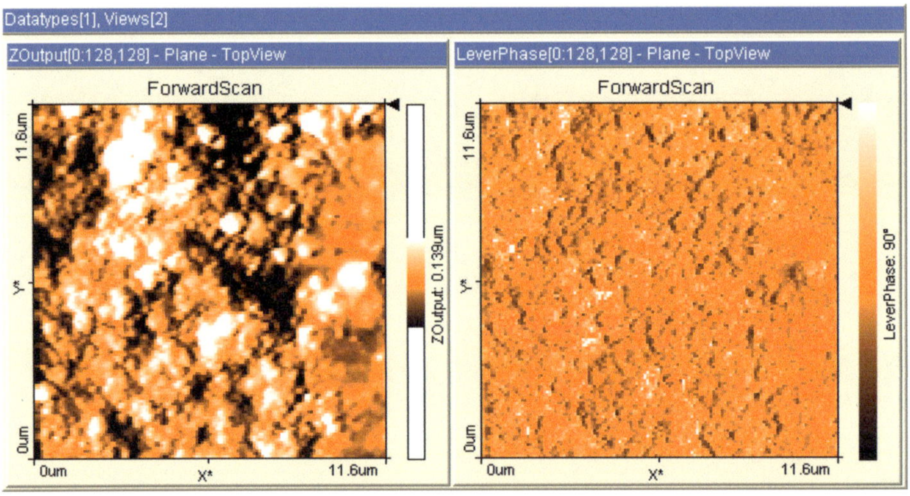

Figure 7.
AFM surface images composite № 5 (RHAPP nano, TESPT). Scans 11.6 × 11.6 μm². Left - topography, right - phase contrast.

The topography and phase contrast image shows that the filler ultrasonic treatment prior to introduction into the elastomer failed to ensure a uniform size distribution of the filler in the rubber matrix as in composite No. 2. This is definitely due to a different chemical composition, which is not favorable for this filler effective interaction with the rubber matrix. From the data of **Figures 2–9** and the results of SEM and AFM images we can see that reducing the filler particles size leads to a marked decrease in the size of inclusions (particle agglomerates) and using the TESPT leads a better filler particles distribution in the elastomeric matrix volume. These results are important for understanding the synthesized composites physical-mechanical properties. The elastomeric composites stress – strain properties are presented in **Figure 10**.

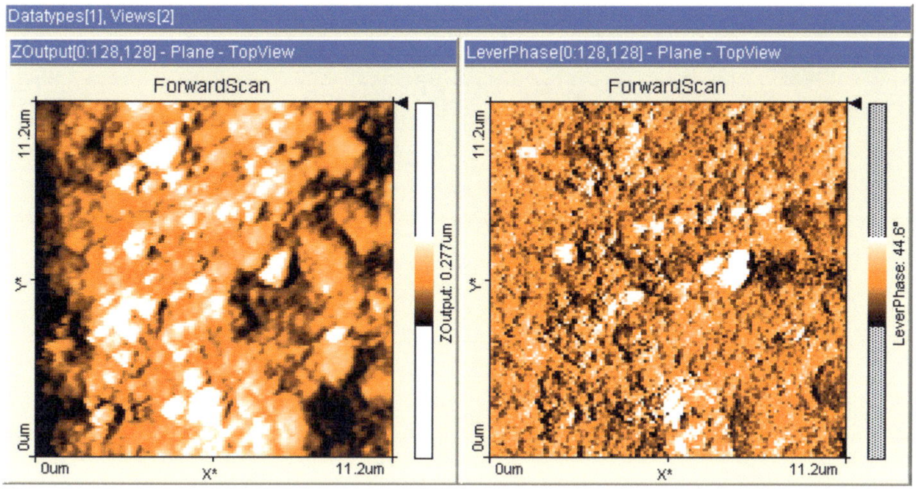

Figure 8.
AFM surface images. Composite № 6(RHAPP nano, oxidized/TESPT). Scans 11.5 × 11.2 μm². Left - topography, right - phase contrast.

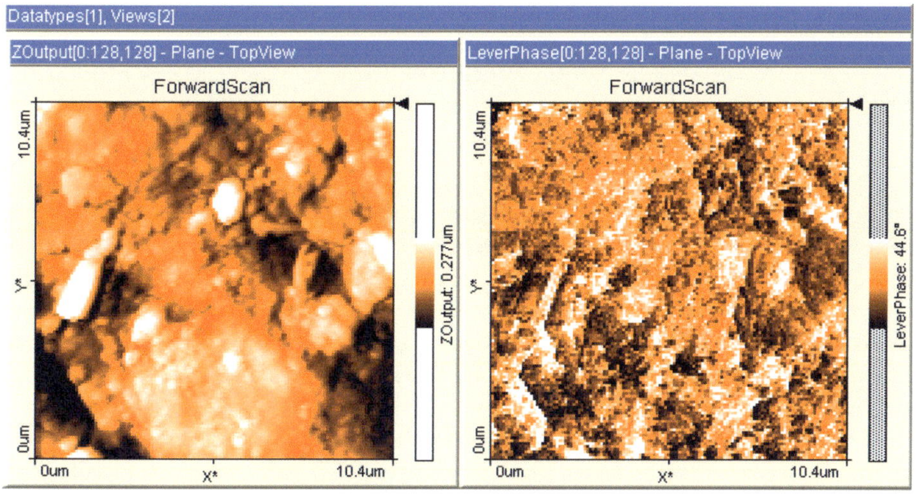

Figure 9.
AFM surface images composite № 7 (RHAPP black micro US). Scans 10.4 × 10.4 μm². Left - topography, right - phase contrast.

For composites with micro particles, average strength values were obtained at the level no more than 4.0 MPa regardless of the fillers composition (samples № 2 (RHAPP micro US) and № 7 (RHAPP black micro US), therefore not presented in **Figure 10**. For composites with submicron particles the average strength reaches about 20 MPa: samples № 3 (RHAPP nano), № 4 (RHAPP nano H_2O_2), № 5 (RHAPP nano TESPT), № 6 (RHAPP nano H_2O_2/TESPT). For samples № 3 (RHAPP nano) there is a significant increase in composites elongation more than 700%. Oxidation of the submicron particle surface leads to a slight decrease in composite strength and elongation - sample № 4 (RHAPP nano H_2O_2). The TESPT combination agent addition to compositions with RHAPP submicron particles - samples № 5 (RHAPP nano TESPT) and № 6

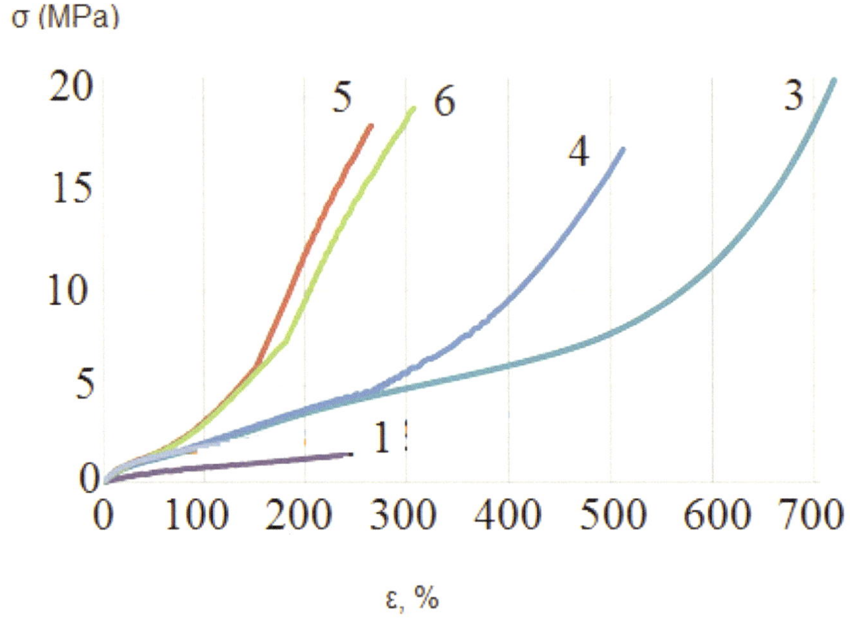

Figure 10.
Engineering stress σ (MPa) – Strain ε (%) curves for elastomeric composites samples: № 1 (pure SBR-30ARK), № 3 (RHAPP nano), № 4 (RHAPP nano TESPT), № 5 (RHAPP nano H_2O_2/TESPT), № 6 (RHAPP nano H_2O_2/ TESPT).

(RHAPP nano H_2O_2/TESPT), mainly leads to a decrease in elongation at break. and a significant increase in elongation stresses up to 20.0 MPa. The composite with an oxidized filler surface (sample № 6) is distinguished by a slightly higher relative elongation and somewhat higher strength. The coupling agent use in the case of non-oxidized submicron particles RHPP on the contrary, leads to a slight decrease in strength (comparison of samples № 3 and № 5). These effects show a significant effect of the surface functionality of submicron filler particles on the elastomeric composites mechanical properties with rice husk processing products. The strength values obtained in this work for samples № 3 (RHAPP nano), № 4 (RHAPP nano H_2O_2), № 5 (RHAPP nano TESPT), № 6 (RHAPP nano H_2O_2/TESPT) exceeding 20 MPa, are considerably higher than values about 13 MPa obtained in other investigations [23, 28, 30].

The mechanical behavior of elastomeric composites under dynamic loading conditions was further investigated. The elasticity dynamic modulus dependence on the amplitude of deformation (**Figure 11**) was obtained on a rheometer HAAKE RheoStress RS 150. This study was carried out to evaluate the Payne effect, which characterizes the interaction strength of filler particles with each other, as well as its change depending on the fillers crushing factor. The frequency was 0.5 Hz and the temperature 60°C. Tests were carried out on a rheometer in plane-plane mode, the amplitude of deformation is represented on the x-axis, and the y-axis represents the complex dynamic modulus. It can be seen from the data that the complex dynamic modulus value is significantly higher for sample № 3 (RHAPP nano) at all strain amplitudes. This indicates a strong particle-particle interaction (filler agglomeration) for sample № 3 (RHAPP nano) filled with submicron particles. For sample № 2 (RHAPP micro), these effects are significantly less pronounced due to the larger size of the inclusions.

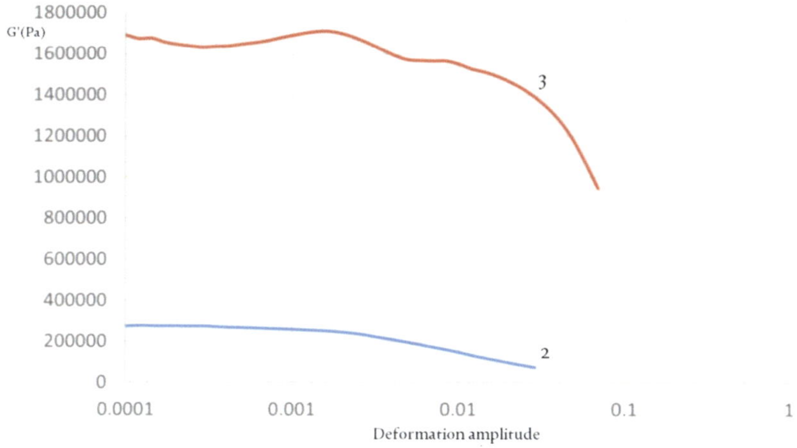

Figure 11.
The real component of complex dynamic modulus G'(Pa) dependence on deformation amplitude for elastomeric composites: (sample № 2) micro-dispersed filler; (sample № 3) highly dispersed filler with submicron particles.

Thus, significant enhancement of physical and mechanical properties of composites of RAS rubber with rice husk ash fillers achieved in this study was possible due to two important components of composite fabrication. The first important component, as shown by synthesized composites SEM and AFM studies of the surface structure, is the filler dimensions transfer into the nanoscale. The second important component is the filler surface chemical modification to provide better adhesion to the elastomer matrix. SBR rubber matrix is hydrophobic while amorphous silica derived from rice husk ash is hydrophilic. In order to ensure their effective interaction and, consequently, a significant enhancement of physical and mechanical properties, it is necessary to use silanol-based adhesives, one of which is TESPT. Scheme of interaction of hydroxyls of amorphous silicon dioxide surface with TESPT is shown in **Figure 12** [31]. Alkoxyl groups of TESPT interact with hydroxyls of the filler, whereas dissociating sulfur groups interact with the elastomeric matrix. This leads

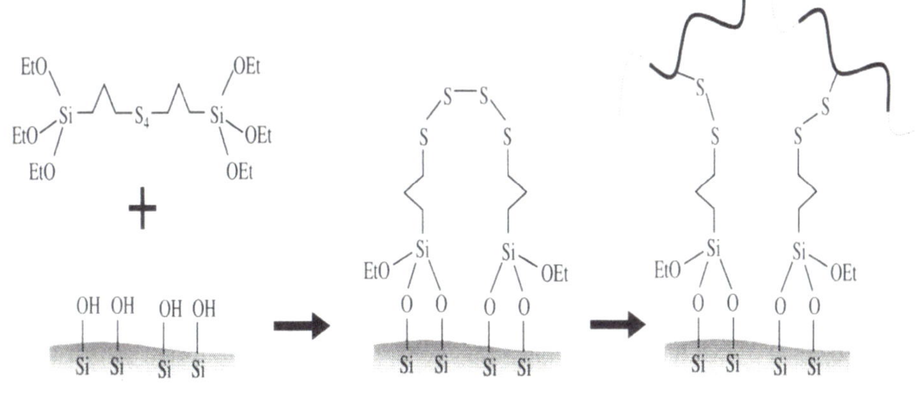

Figure 12.
TESPT interaction mechanism with the silica surface [31].

to a synergistic interaction of the inorganic filler with the rubber, resulting in the composites enhanced deformation and strength properties.

4. Conclusions

The elastomeric composites study by electron and atomic force microscopy methods has been shown the influences of surface structure parameters of elastomeric composites on the complex of their mechanical properties. The directional control effectiveness over the initial particles size of amorphous silicon dioxide derived from renewable raw materials (rice husk ash) was shown. The Chapter demonstrates the efficiency of surface functionalization amorphous silicon dioxide particles and these particles application as promising reinforcing filler for composites with a complex of high elastic strength properties. In the proposed work, submicron particles of amorphous silicon dioxide were obtained for the first time by ultrafine grinding of rice husk ash processing products. Also for the first time the great effect of reinforcing with mineral filler from rice husk ash processing products for elastomeric composites based on rubber SBR-30 ARK was established. The relationship of the structure and complex of mechanical properties for elastomeric composites with a new class of strengthening mineral fillers has been established. For a sample filled with rice husk processing products with a surface modifier, the average strength value was obtained at the level 20 MPa, which was achieved for the first time for elastomeric composites with rice husk processing products as filler. It is shown that when grinding the rice husk ash particles by the method developed at IAM RAS and using them as filler, an elastomeric composite was obtained, approaching the elastic-strength properties of samples with precipitated silica as filler in the present rubber industry. The data obtained indicates the relationship between the size of filler particles and their reinforcing effect in the elastomeric composite. This dependence is known for fillers such as carbon particles and silica. It is shown that this dependence is also performed for mineral fillers of natural origin – rice husk ash processing products.

Acknowledgements

Work was done according the State Research Program for Institute Applied Mechanics Russian Academy of Sciences (IPRIM RAS).

Conflict of interest

The authors declare no conflict of interest.

Author details

Yury V. Kornev[1], Hammat H. Valiev[1*], Yuliya N. Karnet[1], Alla A. Kornilova[2], Nikolay A. Semenov[1] and Alexander N. Vlasov[1]

1 Institute of Applied Mechanics Russian Academy Sciences, Moscow, Russia

2 Physics Faculty, Moscow State University, Moscow, Russia

*Address all correspondence to: hhvlv@mail.ru

References

[1] Pereira AB, Fernandes FAO, editors. Renewable and Sustainable Composites. London: IntechOpen; 2019. p. 139. DOI: 10.5772/intechopen.77412

[2] Saleh HEM, Koller M, editors. Characterizations of some Composite Materials. London: IntechOpen; 2019. p. 152. DOI: 10.5772/intechopen.80960

[3] Yanovsky Y. Nanomechanics and Strength of Composite Materials. Moscow: IPRIM RAS; 2008 179 p

[4] Bakar R, Yahya R, Gan S. Production of high purity amorphous silica from rice husk. Procedia Chemistry. 2016;**19**:189-195. DOI: 10.1016/j.proche.2016.03.092

[5] Liu X, Chen X, Yang L. Mint: A review on recent advances in the comprehensive application of rice husk ash. Research on Chemical Intermediates. 2016;**42**:893-913. DOI: 10.1007/s11164-015-2061-y

[6] Soltani N, Bahrami A, Pech-Canul M, González L. Review on the physicochemical treatments of rice husk for production of advanced materials. Chemical Engineering Journal. 2015;**264**:899-935. DOI: 10.1016/j.cej.2014.11.056

[7] Kenechi N-O, Linus C, Kayode A. Utilization of rice husk as reinforcement in plastic composites fabrication – A review. American Journal of Materials Synthesis and Processing. 2016;**1**(3):32-36. DOI: 10.11648/j.ajmsp.20160103.12

[8] Zou Y, Yang T. Rice husk, rice husk ash and their applications. In: Rice Bran and Rice Bran Oil. AOCS Press, Elsevier; 2019. pp. 207-246. DOI: 10.1016/B978-0-12-812828-2.00009-3

[9] Soltani N, Bahrami A, González L. Review on the physicochemical treatments of rice husk for production of advanced materials. Chemical Engineering Journal. 2015;**264**:899-935. DOI: 10.1016/j.cej.2014.11.056

[10] Rout K, Satapathy A. Development and characterization of rice husk-filled glass fiber-reinforced epoxy biocomposites. Composites: Mechanics, Computations, Applications: An International Journal. 2012;**3**:95-106. DOI: 10.1615/CompMechComputApplIntJ.v3.i2.10

[11] Arjmandi R, Hassan A, Majeed K, Zakaria Z. Rice husk filled polymer composites. International Journal of Polymer Science. 2015;**2015**:501471. DOI: 10.1155/2015/501471

[12] Bilal A, Lin R, Jayaraman K. Optimal formulation of rice husk reinforced polyethylene composites for mechanical performance: A mixture design approach. Journal of Applied Polymer Science. 2014;**131**:1-10. DOI: 10.1002/app.40647

[13] Bilal A, Lin R, Jayaraman K. Optimisation of material compositions for flammability characteristics in rice husk/polyethylene composites. Journal of Reinforced Plastics and Composites. 2014;**33**:2021-2033. DOI: 10.1177/0731684414552542

[14] Chanda A, Neogi S, Neogi S. Optimization of plasma treatment for enhanced filler matrix adhesion in manufacturing green composites with rice husk. Indian Chemical Engineer. 2013;**55**:177-188. DOI: 10.1080/00194506.2013.832026

[15] Chanda A, Hazra A, Kumar M, Neogi S, Neogi S. Chemical treatments of rice husk filler and jute fiber for

the use in green composites. Fibers and Polymers. 2015;**16**:902-910. DOI: 10.1007/s12221-015-0902-3

[16] Nishata R, Sulong A, Yuliana N, Sahrim A. Effect of surface modified rice husk (RH) on the flexural properties of recycled HDPE/RH composite. Advances in Materials and Process Technologies. 2017;**3**:1-8. DOI: 10.1080/2374068X.2017.1340420

[17] Fernandes MRS, Maria FA, Furtado C. Rice husk ash as filler in tread compounds to improve rolling resistance. Polímeros. 2017;**27**:55-61. DOI: 10.1590/0104-1428.2385

[18] Lemessa Jembere A, Workneh FS. Studies on the synthesis of silica powder from rice husk ash as reinforcement filler in rubber tire tread part: Replacement of commercial precipitated silica. International Journal of Material Science Appliences. 2017;**6**:37. DOI: 10.11648/j.ijmsa.20170601.16

[19] Vishvanathperumal S, Anand G. Effect of nanoclay/nanosilica on the mechanical properties, abrasion and swelling resistance of EPDM/SBR composites. SILICON. 2020;**12**:1925-1941. DOI: 10.1007/s12633-019-00291-6

[20] Sintharm P, Phisalaphong M. Green natural rubber composites reinforced with black/white rice husk ashes: Effects of reinforcing agent on film's mechanical and dielectric properties. Polymers (Basel). 2021;**13**:882. DOI: 10.3390/polym13060882

[21] Boonmee A, Sabsiriroht P, Jarukumjorn K. Preparation and characterization of rice husk ash for using as a filler in natural rubber. Materials Today: Proceedings. 2019;**17**:2097-2103. DOI: 10.1016/j.matpr.2019.06.259

[22] Xiao Y, Zou H, Zhang L, Mint. Surface modification of silica nanoparticles by a polyoxyethylene sorbitan and silane coupling agent to prepare high-performance rubber composites. Polymer Testing. 2020;**81**:106195. DOI: 10.1016/j.polymertesting.2019.106195

[23] Bach Q, Vu C, Vu H. Effects of co-silanized silica on the mechanical properties and thermal characteristics of natural rubber/styrene-butadiene rubber blend. SILICON. 2020;**12**:1799-1809. DOI: 10.1007/s12633-019-00281-8

[24] Dominic M, Joseph R, Sabura Begum P, Mint. Green tire technology: Effect of rice husk derived nanocellulose (RHNC) in replacing carbon black (CB) in natural rubber (NR) compounding. Carbohydrate Polymers. 2020;**230**:115-620. DOI: 10.1016/j.carbpol.2019.115620

[25] Kornev Y, Semenov N, Vlasov A, Valiev K. Reinforcing effects in elastomeric composites, filled with particles of mineral fillers, based on silicon dioxide and carbon. Journal of Physics: Conference Series. 2021;**1942**:012031. DOI: 10.1088/1742-6596/1942/1/012031

[26] Valiev H, Vlasov A, Kornev Y, Mint. Scanning probe microscopy of elastomers with mineral fillers. In: Pereira A, Fernandes F, editors. Renewable and Sustainable Composites. London: IntechOpen; 2019. pp. 39-50. DOI: 10.5772/intechopen.84554

[27] Garishin O, Shadrin V, Kornev Y. Mechanical studies of rubber micro and nanocomposites promising for the tire industry. Uniaxial and biaxial tests. Materials Physics and Mechanics. 2019;**42**:445-454. DOI: 10.18720/MPM.4242019_9

[28] Vilmin F, Bottero I, Travert A. Mint: Reactivity of bis[3-(triethoxysilyl) propyl] tetrasulfide (TESPT) silane

coupling agent over hydrated silica: Operando IR spectroscopy and chemometrics study. Journal of Physical Chemistry C. 2014;**118**:4056-4071. DOI: 10.1021/jp408600h

[29] ASTM D412-16. Standard Test Methods for Vulcanized Rubber and Thermoplastic Elastomers—Tension. West Conshohocken, PA: ASTM International; 2021. DOI: 10.1520/D0412-16

[30] Ulfah I, Fidyaningsih R, Rahayu S, Mint. Influence of carbon black and silica filler on the rheological and mechanical properties of natural rubber compound. Proceedings of Chemical Society. 2015;**16**:258-264. DOI: 10.1016/j.proche.2015.12.053

[31] Donnet J, Custodero E. Reinforcement of elastomers by particulate fillers. In: Mark JE, Erman B, Roland M, editors. The Science and Technology of Rubber. 4th ed. Amsterdam: Elsevier; 2013. pp. 383-416. DOI: 10.1016/B978-0-12-394584-6.00008-X

Chapter 4

Sugar Cane Bagasse Ash: An Agricultural Residue with Potential Rubber Filler Applications

Ntalane S. Seroka, Raymond Taziwa and Lindiwe Khotseng

Abstract

South Africa produces approximately 7 million tons of sugarcane bagasse annually as an agricultural residue, which is treated as waste and its disposal is known to have negative impacts on the environment. To lessen reliance on petroleum and polymers, consideration is given on use of sugarcane bagasse ash as substitute materials for the development of fillers for rubber and other large-scale commodity polymers. This work reports on the mechanical, physiochemical, and structural properties of sugarcane bagasse ash to define the compatibility with the specific polymers that will pave way to the engineering of composites to utilize the potential benefits of these residue-derived fillers. The structural and morphological properties of the untreated and treated sugarcane bagasse ash were performed using XRD, FTIR, and SEM-EDX, respectively. The obtained results confirmed the successful treatment of the sugarcane bagasse ash. The study was successful in showing that sugarcane bagasse ash as potential filler in rubber polymer matrix is a natural resource of silica, which is sustainable and cost-effective, thus should be harnessed for industrial purposes in South Africa.

Keywords: sugarcane bagasse ash(SCBA), fillers, rubber materials, composites, eco-friendly

1. Introduction

Sugarcane commonly known as "*saccharum officinarum L*" is a commercially grown crop in harvesting season under climate conditions of slightly sunny and colder. And, it is produced annually in most parts worldwide, with 100 million tonnes produced by both Brazil and India. South Africa produces about 19.3 million tonnes of sugarcane annually followed by Nigeria and Uganda with two of largest producers of sugar cane on African soil [1–3].

Sugar cane in South Africa is grown in the Kwa-Zulu Natal Province and Mpumalanga Province as shown in **Figure 1**. Sugar cane production in the Kwa-Zulu Natal Province is composed of gross farming located across the two provinces whereby about 22,949 registered sugarcane growers produce annually 20 million tons of sugarcane from approximately 14 million supply sections, extending from southern Kwazulu-Natal to Mpumalanga Lowveld as shown in **Figure 1** [4].

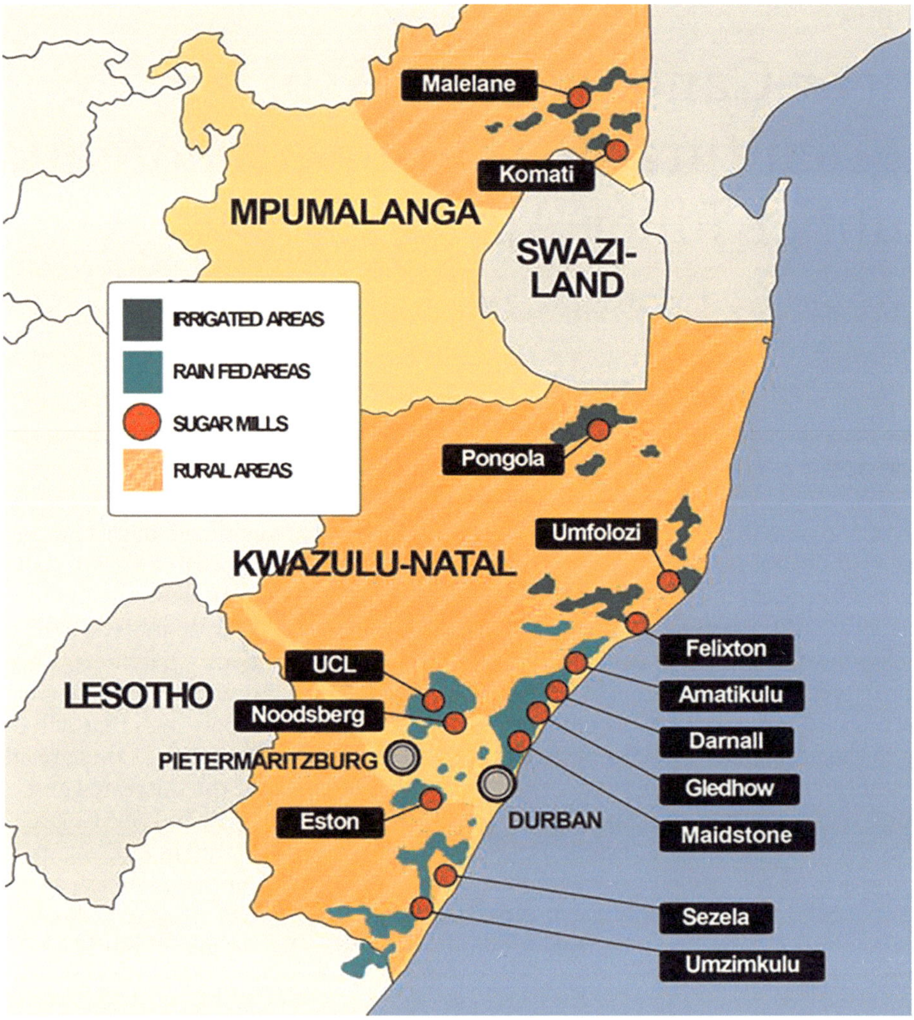

Figure 1.
Represents sugarcane plantations regions in the Mpumalanga and KwaZulu-Natal provinces of South Africa.

Sugarcane primarily it is the source of sugar and juice, and can be used as a raw material in various industries for production of sugar, jiggery and syrups. In spite of that these industrial processes generate considerable quantities of residual bagasse, which is often used to produce heat energy via combustion. Interestingly, about 25–30 wt residual lignocellulose bagasse (RSB) is produced per kg of sugarcane processed in sugar mills [5–8]. The crop accumulates to approximately 380 kg/pa of silicon in a 12-month-old-crop. In addition, silicon in this state is a solid waste as silica (SiO_2). It has been reported that natural silica is safe to handle, cheap and simple work-up extraction from various sources. The fate of bagasse ash poses a challenge to the environment [9].

Nowadays, research has swiftly shifted focus to the industrial agricultural wastes to address the ever-growing concerns for the inadequate disposal of residues produced from agro-wastes. Consequently, led to the development of hybrid composites via reinforcements in polymer science. The use of substitute materials such as sugarcane bagasse as a filler in rubber and cementious application, has become a global concern to look at ways to minimize the disposal of wastes especially those which are non-biodegradable and pose a threat to the eco-system. Therefore, recycling and re-use of these residues released as wastes from industries has been efficient route to minimize the disposal of wastes which are not environmentally friendly and pose a threat to the quality of life (micro-organisms) [10].

Notably the re use of recycled fillers as reinforcements from renewable sources has received an increased attention with the aim of synthesizing alternate materials in addressing shortcomings related to the fairly and low sustainable of conventionally reinforced polymer composites. Carbonaceous nanomaterials from various sources such as wood, jute, cotton and bagasse have been introduced into a polymer matrix as reinforcements which have recently gained momentum in polymeric composites, this is due to their production means being inexpensive, fast and do-able [11].

Additionally, sugarcane bagasse their composites possess good chemical, physical and mechanical properties and is considered a low-cost and low-density material which possesses potential application-specific mechanical strength and stiffness values [12]. Thus, researchers reported on the mechanical properties of the composites (elasticity and elongation at break) whereby they measured using the average of the three properties in **Table 1**. In relation to the stress–strain, the mechanical properties in terms of tensile strength, the values measured from sugarcane bagasse ash SCBA with optimized silane treatment were found to be the most desired and excellent.

For modified application of silane coupling agent, **Bis[3-(triethoxysilyl)propyl] Tetrasulfide (TESPT)**, in the SCBA-Natural Rubber interaction, SCBA particles were homogenously dispersed in the NR matrix, from the method reported in [13, 14]. The samples with a significant amount (phr) of SCBA were studied to understand and aim for greater interfacial adhesion within the treatments. Hence, the composites displayed improved mechanical response, the bold in the table imply to the best results observed in the samples. Moreover, the use of SCBA can be a feasible substitute material as filler in Natural Rubber (NR) due to excellent mechanical properties [14].

Interestingly, sugarcane bagasse has been used as reinforcing phase in cementitious and polymer composites. And, it comprises of waste generated from industrial production of sugar, fuel and other beverages derived from sugarcane. Sugarcane bagasse ash is mostly composed of silica in various forms structurally including crystalline, vitreous and amorphous. The amorphous silica from SCBA has been utilized as a filler in polymer composites. Wimolmala and Sombatsompop reported for the first time the utilization of SCBA with a particle size in the range 45–145 μm as a filler in a rubber matrix with a concentration of 15 phr (per hundred rubber). Although, they incorporated high content of N-cyclihexyl-2-benzothiazole sulphonamide (CBS), which, in spite of improving the crosslinks sections and the mechanical resistance of

Property	Pure gum	25 per hundred rubber (phr) pristine	25 per hundred rubber (phr) in situ silane treatment	25 per hundred rubber (phr) with optimized silane treatment
Stress (MPa)	4.02	4.23	4.96	**11.47**
Modulus (E) (MPa)	0.32	0.37	1.03	0.62
Strain (%)	974	625	442	**857.4**

Table 1.
Mechanical properties of the composites with and with no silane coupling agent.

the composites, significantly minimized their life time. Traditionally, the alkaline treatment is utilized followed by a strong acid for neutralization to produce silica gel, as illustrated below [13–15]:

$$SiO_2(s) + 2NaOH_{(aq)} \rightarrow Na_2SiO_{3(aq)} + H_2O_{(aq)} \quad (1)$$

$$Na_2SiO_{3(aq)} + 2HCl_{(aq)} \rightarrow SiO_{2(gel)} + 2NaCl_{(aq)} + H_2O(l) \quad (2)$$

Other studies have reported the use of sugarcane fibers as reinforcing filler, which implies the environmentally-friendly application of this natural resource. In this current study, the structural and morphological properties of sugarcane bagasse ash were studied as well as thermal properties potential for reinforcements in rubber materials. The use of citric acid (an organic acid) during treatment step to remove any inorganic impurities from the ash proved to be useful tool in realizing green methodology for the use of sugarcane bagasse ash [16].

Nanomaterials have allowed researchers to effectively manipulate and exploit materials at the nanometer scale to produce new functionalities. And, to fine-tune their properties from an atomic level, to get desired properties to suit particular applications. In essence, rubber materials are desired to be inexpensive, due to the biomaterials (filler/polymer matrix) and their fabrication methods as well as their resources which is of paramount importance as compared to conventional silica from non-natural resources. In this work surface treatment was performed using citric acid potentially to improve the interfacial adhesion of waste and its higher interactions with elastomeric matrix. The structural, morphological and thermal properties of SCBA potential for filler matrix interaction were analyzed.

2. Materials and methods

2.1 Material used

The chemicals utilized in the preparation were citric acid ≥99.5%, purchased from Sigma Aldrich. Sugarcane was procured from Sugar Illovo South Africa Company. The synthesis was done using deionized water from the Milli-Q water purification system (Millipore, Bedford, MA, USA).

2.2 Preparation of Sugarcane Bagasse Ash (SCBA)

The sugarcane bagasse (a, b) cultivated around the tropical and coastal regions of South Africa is used in this work, shown in **Figure 2**. Firstly, the incomplete combustion of the bagasse to ash (c, d) was carried out in open space and then heated in an oven at 40°C overnight to obtain uniform particle sizes from dried ash. In a typical procedure the residue of (d) sugarcane bagasse 2.0 g heated at 700°C for 4 hours (residence time) in a muffle furnace at 10°C/min heating rate (gradient time) to burn away volatile organic matter illustrated in **Figure 2D**.

2.3 Leaching with citric acid

In the first treatment, the bagasse ash was mixed with citric acid in 250 ml beaker and then underwent reflux in a 250 ml volumetric flask stirred at 450 revolution per minute (rpm) at 80°C for 2 hours, shown in **Figure 2D, E**. The ash was then washed

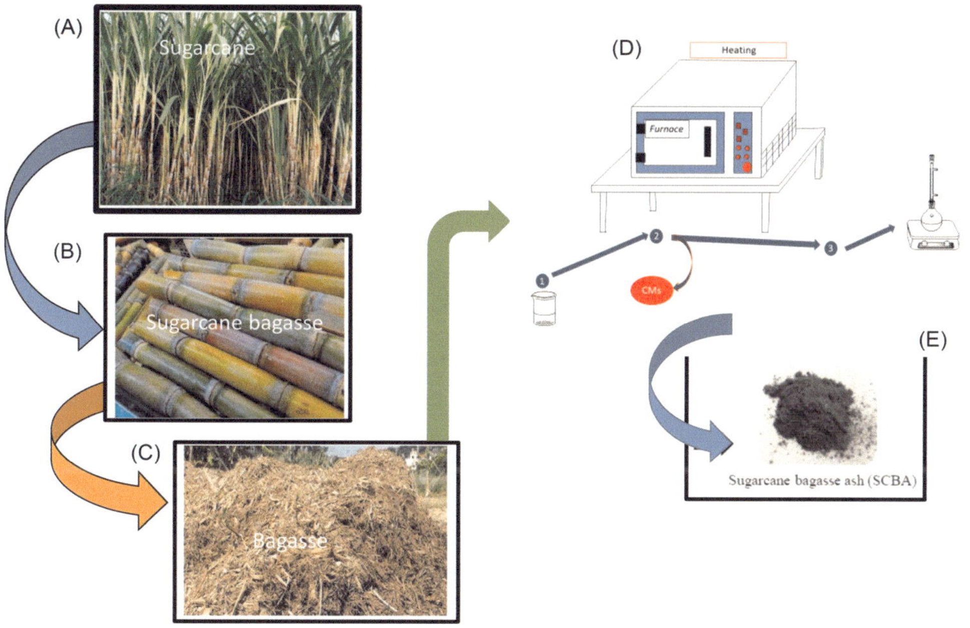

Figure 2.
Presents images of a typical sugarcane plantation (A), stalks (B), and bagasse (C) from the south African sugar industry, calcination in a muffle furnace (D), and conversion to sugarcane bagasse ash (E).

with double deionized water and decant until the pH of the supernatant reached 6.5. The resulting ash was then dried in an oven at 40°C overnight and ground into fine powder using mortar and pestle.

3. Results and discussion

3.1. Structural and morphological characterization.

The crystallinity identification was cross-examined using X-ray diffraction (XRD) on a Bruker AXSD8 Advancement instrument, (Ithemba Labs, South Africa) with Cu-Kα1, λ = 154,050 A. The Bragg angle array was 2Θ = 10–90°C with a scanning step 0,035°C. The new surface functionalities were studied using IR by identifying the functional groups and bonding of elements present in the samples. The analysis was undertaken at room temperature with a wavelength range 400–4000 cm^{-1} and phase composition determined using a PerkinElmer FTIR spectrometer (spectrum two). Scanning electron microscopy analysis was performed using SEM (TESCAN, VEGA) to observe surface morphology of the sugarcane bagasse ash (SCBA). The samples were prepared on an aluminum stub and carbon sputtered on a carbon coater before analysis.

3.1 Characterization of synthesized materials

3.1.1 Powder X-ray diffraction and FTIR analysis

The X-ray diffraction pattern of untreated (i) and treated (ii) Sugarcane bagasse ash (SCBA) is presented in **Figure 3a**. The x-ray diffraction pattern reveals diffraction peaks associated with the presence of quartz silica @ 2Θ = 27 (JCP phase from ICDD: 01–083-0539), and broad peak corresponding to the amorphous nature of silica in the bagasse sample @ 2Θ = 12 and between 2Θ = 22 and 28. The attributed traces of silica in the form of quartz in the bagasse ash was reported to be as a result of accumulation of minerals from the soil. The obtained bagasse proves to be heavily impure as shown in **Figure 1a** (i) for untreated bagasse ash. The acid treated bagasse, **Figure 1a** (ii) ash show crystalline peaks attributed to silica as well as smooth peaks as compared to the untreated bagasse ash [17].

Figure 3b, presents FTIR spectra of untreated (i) and acid-treated (ii) bagasse ash. The key functional groups existing in the bagasse ash were identified and characteristics peaks for SCBA with absorption bands between 1046 and 1221 cm^{-1} corresponds to asymmetric vibration of the Si-O-Si bonds. The peaks at 464 and 775, correspond to Si-OH present on the surface of the particles. The peaks at 1361 and 1740 are associated with organic matter in the material and C=O, respectively. The peaks at 3014 and 3457 are as a result of organic matter in the ash C-O and OH group, respectively.

SCBA material it is possible to observe the pronounced peaks and intense peaks for the acid leached sample in **Figure 3b** (ii). It is noteworthy to realize the evolution and disappearance of functional groups and the appearance of key functional groups. This confirms successful incorporation of new functionalities on the acid leached sample (ii) **Figure 3b**. Moreover, the highly narrow and pronounced band at around 1046 cm^{-1} which is associated with crystalline silica, is in agreement with the results obtained from XRD diffractograms(s) [17, 18].

3.1.2 SEM–EDX analysis

The surface morphologies for the untreated and treated ash were cross examined using Scanning Electron Microscopy (SEM), presented in **Figure 4A** untreated bagasse and (c) for acid-treated bagasse ash. The SEM images, it is possible to observe the agglomerates of irregular morphology and sizes as well as spherical-rice like particles present in the samples. Although, organic acid leaching process was effective on the surface morphology as observed in (d) with rice-like shape observed, the morphological differences confirm the surface characteristics analysis (FTIR) discussed earlier, where the pretreatment

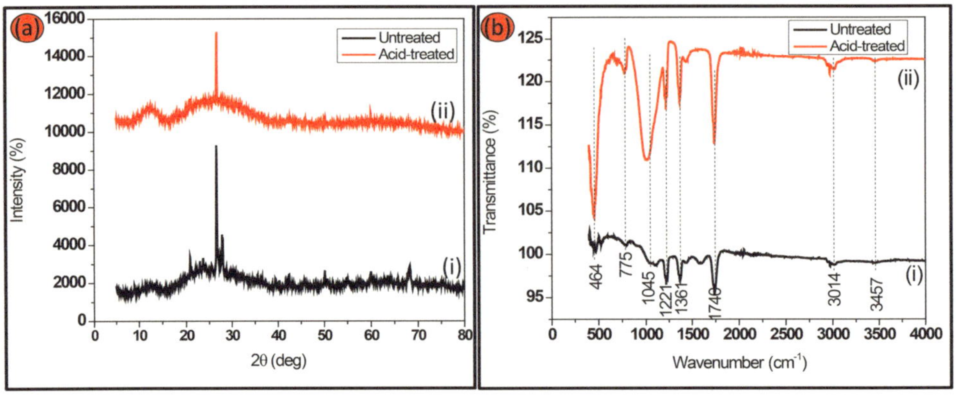

Figure 3.
Presents XRD spectra of (a) untreated bagasse ash (i) and treated bagasse ash with citric acid (ii), and FTIR spectra of (b) untreated bagasse ash (i) and treated bagasse ash with citric acid (ii).

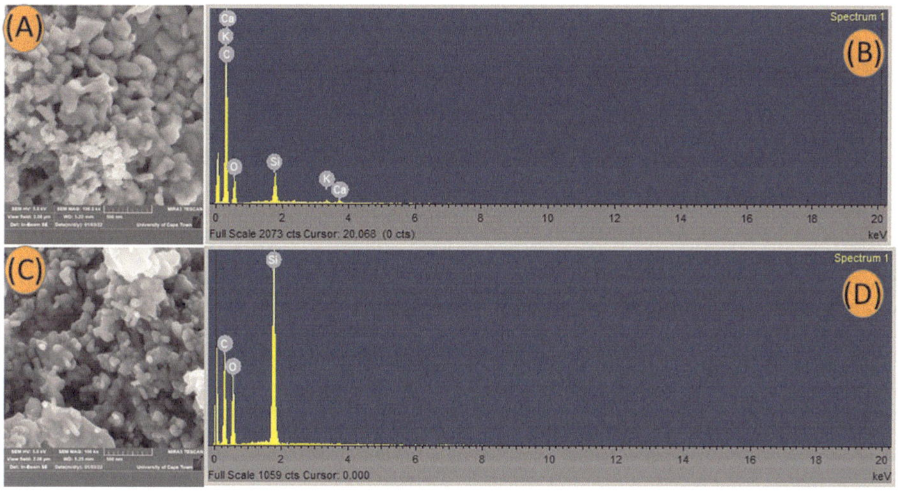

Figure 4.
SEM images of (A) untreated bagasse ash, corresponding EDX spectrum (B) and image of (C) treated bagasse ash with citric acid and corresponding EDX spectrum (D).

with citric acid significantly improved the formation of silica for the acid-leached sample [18, 19].

The EDX results exhibit that the chemical composition of SCBA untreated and treated with citric acid obtained by SEM–EDX shown in **Figure 4B** and **D**, respectively, that acid treatment is an effective tool to mitigate impurities from the increment of Si matter in the samples for acid-leached. In addition, the acid-treated bagasse ash reveal closely packed particles due to the increased formation of hemicellulose removal bonds. The acid leaching treatment results in the degradation of cellulose and hemicellulose-inspired bonds, thus the resulting cellulose chains generate new hydrogen bonds to replace the removed ones. In essence, improves filler polymer interaction [20].

Table 2 shows that citric acid is a useful organic acid for the pretreatment process. The increased morphology of silica in the powders of the ash is as result of the selective removal of the synthesis residues using citric acid, thereby reducing metallic impurities present as compared to the morphology from the untreated ash.

The samples show that pretreatment of the ash results in significant reduction in carbon content as shown in **Table 1** above, and silicon increased from 1.78% to 5.27%.

It is quite evident that the pretreatment step is a useful tool and sufficient in purifying the bagasse ash by reducing inorganic impurities to achieve high yield of silica implied by the mathematical expressions below [21]

$$\text{100 g of SBA introduces approximately 76.34 g of } SiO_2 \tag{3}$$

$$\text{Therefore,unknown (X) amount of SCBA will introduce 73.21 g of } SiO_2 \tag{4}$$

$$\text{The aforementioned implies,} = \frac{100\ x\ 73}{76.34} = \frac{7300}{76.34} = 95.625\, gofSCBA \tag{5}$$

For, Al_2O_3 since 100 g SCBA introduces 6.7 g of Al_2O_3,

$$\text{then : 95.625 g of SCBA will introduce} = \frac{95.625\, x\ 6.7}{100} = \frac{640.688}{100} = 6.407\, gofAl_2O_3 \tag{6}$$

Element	Untreated SCBA (%)	Acid-treated SCBA (%)
C	74.92	69.36
O	22.45	25.37
Si	1.78	5.27
K	0.26	—
Ca	0.59	—

Table 2.
Chemical composition of raw bagasse ash.

$$Fe_2O_3 : \text{Since 100 g SCBA introduces 6.3 g of } Fe_2O_3,$$
$$\text{then} : \text{95.625 g of SCBA will introduce} \frac{95.625\,x\,6.3}{100} = \frac{602.438}{100} = 6.024\,gofFe_2O_3 \quad (7)$$

$$CaO : \text{Since 100 g SCBA introduces 2.8 g } CaO,$$
$$\text{then} : \text{95.625 g of SCBA will introduce} = \frac{95.625\,x\,2.8}{100} = \frac{267.75}{100} = 2.678\,gofCaO \quad (8)$$

$$MgO : \text{Since 100 g SCBA introduces 3.2 g } MgO,$$
$$\text{then} : \text{95.625 g of SCBA will introduce} = \frac{95.625\,x\,3.2}{100} = \frac{306}{100} = 3.06\,gofMgO \quad (9)$$

$$P_2O_5 : \text{Since 100 g SCBA introduces 4.0 g } P_2O_5,$$
$$\text{then} : \text{95.625 g of SCBA will introduce} = \frac{95.625\,x\,4.0}{100} = \frac{382.5}{100} = 3.825\,gofP_2O_5 \quad (10)$$

$$Na_2O : \text{Since 100 g SCBA introduces 1.1 g } Na_2O,$$
$$\text{then} : \text{95.625 g of SCBA will introduce} = \frac{95.625\,x\,1.1}{100} = \frac{105.188}{100} = 1.052\,gofNa_2O \quad (11)$$

$$K_2O : \text{Since 100 g SCBA introduces 2.4 g } K_2O,$$
$$\text{then} : \text{95.625 g of SCBA will introduce} = \frac{95.625\,x\,2.4}{100} = \frac{229.5}{100} = 2.295\,gofK_2O \quad (12)$$

$$\text{LOI} : \text{Since 100 g SCBA introduces 0.9 g } LOI,$$
$$\text{then} : \text{95.625 g of SCBA will introduce} = \frac{95.625\,x\,0.9}{100} = \frac{86.063}{100} = 0.861\,gofLOI \quad (13)$$

3.1.3 Thermal analysis

Thermal properties were investigated using thermogravimetric analysis (TGA) from room temperature to 800°C, reported in [22]. The current study gave insightful information on the thermal characteristics of bagasse ash as shown in **Figure 5**. The TG curves essentially showed the total mass degradation of **Figure 5a** 12% and

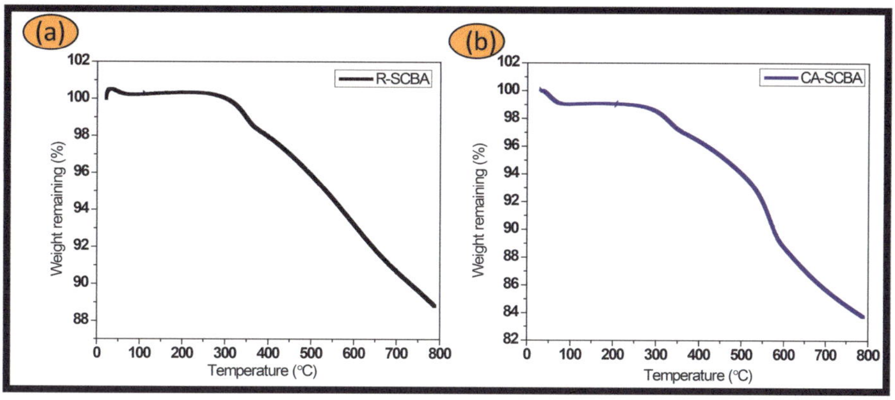

Figure 5.
TGA curves of (a) untreated sugarcane bagasse ash and treated bagasse ash with citric acid. Reproduced with permission from [22], copyright (2022), MDPI.

Figure 5b 18%, respectively. It is noteworthy to realize 3 distinct stages of rapid weight loss attributed to moisture drying, volatile organic matter and rapid decomposition of hemicellulose and cellulose. The results are similar to the findings reported in [23, 24]. No further mass loss was observed as a result of thermal stability of nanosilica present in the samples, which account to 84% in mass, as reported in [25].

4. Conclusions

Based on the findings, we concluded that acid-leaching significantly improved the chemical, physical and thermal properties of bagasse ash. The ash is mostly carbonaceous material. And, the major component of SCBA is silica from the elemental composition studies. The SCBA contains both amorphous and crystalline silica. The reduction of inorganic impurities makes the bagasse residue potential to serve as a filler in polymer composites, and presents a promising and feasible alternative substitute material for the rubber industry. The advantage of this finding is that the use of organic acid proved essentially and equally compatible for the current study as eco-friendly and green method for acid-leaching process.

Furthermore, the filler material will enhance the thermal and mechanical properties of the polymer matrix. These new route to recycle the agricultural residue from the sugarcane industry is pre-conceived as eco-friendly and cost-effective due to the readily available of the plant in tropical countries, and could be used as reinforcing filler in polymer matrix interaction. Therefore, presents a niche application in south Africa as it has never been done before.

Acknowledgements

Funding: This research was funded by National Research Foundation (NRF), South Africa, grant number: 138079 and Tertiary Education Support Program (TESP), Eskom Holdings SOC Limited Reg No 2002/015527/06.

Conflict of interest

The authors declare no conflict of interest.

Author details

Ntalane S. Seroka[1*], Raymond Taziwa[2] and Lindiwe Khotseng[1]

1 Department of Chemistry, University of the Western Cape, Bellville, South Africa

2 Faculty of Science Engineering and Technology, Department of Applied Science, Walter Sisulu University, South Africa

*Address all correspondence to: 3754640@myuwc.ac.za

References

[1] Xu WP, Zhang YY, Wang Q, Li ZJ, Nie YH. Thermoelectric effects in triple quantum dots coupled to a normal and a superconducting leads. Physics Letters A. 2016;**380**(7–8):958-964

[2] Al-Hamamre Z, Saidan M, Hararah M, Rawajfeh K, Alkhasawneh HE, Al-Shannag M. Wastes and biomass materials as sustainable-renewable energy resources for Jordan. Renewable and Sustainable Energy Reviews. 2017;**67**:295-314

[3] Santos F, Rabelo S, De Matos M, Eichler P. Sugarcane biorefinery, technology and perspectives. Philadelphia, Pennsylvania: Academic Press; 2019

[4] Singels A, McFarlane SA, Basdew I, Keeping MG, Nicholson R, Pilusa T, et al. Review of South African sugarcane production in the 2018/19 season: Too much of a good thing? In: Proceedings of the Annual Congress-South African Sugar Technologists Association. South Africa; 2019

[5] Birru E, Erlich C, Martin A. Energy performance comparisons and enhancements in the sugar cane industry. Biomass Conversion and Biorefinery. 2019;**9**(2):267-282

[6] Rabelo SC, Carrere H, Maciel Filho R, Costa AC. Production of bioethanol, methane and heat from sugarcane bagasse in a biorefinery concept. Bioresource Technology. 2011;**102**(17):7887-7895

[7] Canilha L, Chandel AK, et al. Bioconversion of sugarcane biomass into ethanol: An overview about composition, pretreatment methods, detoxification of hydrolysates, enzymatic saccharification, and ethanol fermentation. Journal of Biomedicine & Biotechnology. 2012;**2012**:1-15

[8] Nikodinovic-Runic J, Guzik M, Kenny ST, Babu R, Werker A, Connor KE. Carbon-rich wastes as feedstocks for biodegradable polymer (polyhydroxyalkanoate) production using bacteria. Advances in Applied Microbiology. 2013;**84**:139-200

[9] Norsuraya S, Fazlena H, Norhasyimi R. Sugarcane bagasse as a renewable source of silica to synthesize Santa Barbara Amorphous-15 (SBA-15). Procedia Engineering. 2016; **148**:839-846

[10] Ortiz Zárate D, García-Meca C, Pinilla-Cienfuegos E, Ayúcar JA, Griol A, Bellières L, et al. Green and sustainable manufacture of ultrapure engineered nanomaterials. Nanomaterials. 2020;**10**: 466

[11] Faruk O, Bledzki AK, Fink HP, Sain M. Biocomposites reinforced with natural fibers: 2000–2010. Progress in Polymer Science. 2012;**37**(11): 1552-1596

[12] Mittal G, Rhee KY, Mišković-Stanković V, Hui D. Reinforcements in multi-scale polymer composites: Processing, properties, and applications. Composites Part B: Engineering. 2018; **138**:122-139

[13] Seroka NS, Taziwa RT, Khotseng L. Extraction and synthesis of silicon nanoparticles (SiNPs) from Sugarcane Bagasse Ash: A mini-review. Applied Sciences. 2022;**12**(5):2310

[14] Barrera Torres G, Dognani G, da Silva Agostini DL, dos Santos RJ, Camargo Cabrera F, Gutierrez Aguilar CM, et al. Potential eco-friendly

application of sugarcane bagasse ash in the rubber industry. Waste and Biomass Valorization. 2021;**12**(8):4599-4613

[15] Kaewsakul W. Silica-reinforced natural rubber for low rolling resistance, energy-saving tires: Aspects of mixing, formulation and compatibilization

[16] Katare VD, Madurwar MV. Experimental characterization of sugarcane biomass ash–A review. Construction and Building Materials. 2017;**152**:1-5

[17] Abraham E, Deepa B, Pothan LA, John M, Narine SS, Thomas S, et al. Physicomechanical properties of nanocomposites based on cellulose nanofibre and natural rubber latex. Cellulose. 2013;**20**(1):417-427

[18] Mohd NK, Wee NN, Azmi AA. Green synthesis of silica nanoparticles using sugarcane bagasse. In: AIP Conference Proceedings. Vol. 1885, No. 1. NY, USA: AIP Publishing LLC; 2017. p. 020123

[19] de Paiva FF, de Maria VP, Torres GB, Dognani G, dos Santos RJ, Cabrera FC, et al. Sugarcane bagasse fiber as semi-reinforcement filler in natural rubber composite sandals. Journal of Material Cycles and Waste Management. 2019;**21**(2):326-335

[20] de Lima VM, Barros LC, de Melo Neto AA. Characterization of sugarcane bagasse ash (SBA) and its evaluation for use in alkali-activated slag mixtures. Cerâmica. 2021;**67**:123-130

[21] Plettl A, Enderle F, Saitner M, Manzke A, Pfahler C, Wiedemann S, et al. Non-close-packed crystals from self-assembled polystyrene spheres by isotropic plasma etching: Adding flexibility to colloid lithography. Advanced Functional Materials. 2009; **19**(20):3279-3284

[22] Seroka NS, Taziwa R, Khotseng L. Green synthesis of crystalline Silica from Sugarcane Bagasse Ash: Physico-chemical properties. Nanomaterials. 2022;**12**(13):2184

[23] Yadav AL, Sairam V, Srinivasan K, Muruganandam L. Synthesis and characterization of geopolymer from metakaolin and sugarcane bagasse ash. Construction and Building Materials. 2020;**258**:119231

[24] Rafiee E, Shahebrahimi S, Feyzi M, Shaterzadeh M. Optimization of synthesis and characterization of nanosilica produced from rice husk (a common waste material). International Nano Letters. 2012;**2**(1):1-8

[25] Rafiee E, Shahebrahimi S. Nano silica with high surface area from rice husk as a support for 12-tungstophosphoric acid: An efficient nano catalyst in some organic reactions. Chinese Journal of Catalysis. 2012;**33**(7-8):1326-1333

Chapter 5

Natural Rubber Latex - Origin, Specification and Application

Jacek Rafał Kędzia, Anna Maria Sitko, Józef Tadeusz Haponiuk and Justyna Kucińska Lipka

Abstract

The chapter contains information about the origin of natural rubber latex (NRL) (*Hevea brasiliensis*) and the processing of field latex, considering quality changes occurring during the preparation of raw materials for distribution. The main types of concentrated natural rubber latex are described. A specification of natural rubber latex (NRL) in terms of key parameters tested by manufacturers and customers is presented. Test methods for verifying if the material meets the requirements of ISO 2004 and internal specifications are described based on standards and commonly used techniques. The next subject touched in the chapter is prevulcanization as the processing of concentrated latex with a change of its properties. One of the main industrial applications of NRL as prevulcanized latex (PV) is the production of dipped goods like gloves or balloons. Currently, some trends and challenges relate to sustainability issues are presented (carbon footprint, FSC).

Keywords: natural rubber latex (NRL), field latex, concentrated latex, low ammonia latex, LATZ, high ammonia latex (HA latex), prevucanized latex (PV), natural rubber latex testing methods, dipped goods production

1. Introduction

Natural latex is made from the milk sap of rubber plants. The first historical references to the use of the rubber from the resin of these trees can be found in the relations of Christopher Columbus's expeditions (1492–1496) from Haiti and Fernando Cortez relations from Montezuma's court in Mexico (1519). However, more than 200 years had to pass before Charles Marie de la Condamine presented to the Royal Academy of Sciences in Paris a well-founded description of rubber, the way it was produced, and suggestions for its use with information from C. Franêois Fresneau. Fresneau was probably the first European who described the most productive rubber tree *Hevea brasiliensis* (from Hheve—in the language of the local Indians) [1, 2].

Systematizing the nomenclature: natural rubber (NR) refers to dry rubber, e.g. in sheet, block, or crepe form, while natural rubber latex (NRL) is a suitably stabilized, usually concentrated, liquid form of rubber solution. However, due to the dominance of dry rubber in the market, the term natural rubber (NR) often covers both types of raw materials. Over the years, latex has grown in importance and its production has moved to other parts of the world. Global production of natural rubber (NR and NRL) in year 2018 was 13,8 million of tones, and more than 90% of this production comes from South and South-East Asia [3]. Production of NRL (1,6 million of tones in 2018 [3]) represents approximately 12% of whole natural rubber market. Thailand is the world leader in natural rubber market as a producer of 37% of total rubber and 69% of NRL in 2018 [3].

Unique properties of natural latex as an elastomer makes it applicable in many industries, like the manufacture of dipped products, such as gloves, balloons, condoms, catheters, and other medical products. Foamed latex is used to produce mattresses, pillows, or in footwear industry. Latex is used in adhesive-binders industry or in production of carpets as backing.

Natural rubber latex is a colloidal mixture which means it contains water-based serum phase and the solid rubber phase. The rubber obtained from *H. brasiliensis* is a polymer of cis-1, 4-polyisoprene. Polymer chains have a length from 1000 to 3000 of isoprene units [4].

Latex has very specific organization of the rubber molecules in the rubber particles. Rubber particles are described as spherical or pear shape core-shell structures with polymer core (inside) surrounded by a thin "shell" built of phospholipids and protein. In the model presented by D. C. Blackley [5], there are two layers: inner shell of phospholipids and outer shell of proteins. However, it was also mentioned that proteins and phospholipids are more likely mixed. Mixed structure of the surface has been confirmed and more precisely presented in further works [6]. Each polymer chain has two unique terminal groups: α and ω. α-terminal linked with phosphate groups associated with phospholipids and ω-terminal linked with protein, which play a vital role in forming the unique structures of NR particle [7, 8].

Rubber particles have bimodal diameter size distribution, and rubber particles can be classified based on the size as small rubber particles (SRP) with an average particle size of 10–250 nm and large rubber particles (LRP) with an average particle size of 250–3000 nm [4, 5, 9, 10]. That bimodal size distribution looks different in different types of latices (fresh latex, concentrated latex, and prevulcanized latex (PV) latex) and changes during processing—for example during centrifuging or during clarification [9, 10].

1.1 Stability of latex mixture

Stability of the mixture is the crucial issue in colloidal solution. In case of natural rubber latex, colloidal stability is needed for storage, transport, and processing. Decreasing of this stability brings irreversible aggregation as micro-coagulation and coagulation, but on the other hand, over stabilized latex can be difficult in processing.

From physical point of view, the optimal stability occurs due to the proper interparticle repulsive forces. On the one hand, destabilizing Van der Waals attractive forces act, and on the other hand particles are stabilized by repulsive electrostatic and steric forces. Electrostatic stabilization of latex is consequence of the negative charge of the rubber particle surface with proteins and phospholipids. Around the rubber particle, ions present in the serum solution are adsorbed and create conditions that prevent aggregation. Additional level of protection is a steric stability which is an effect of the presence of adsorbed long-chain molecules at the surface of rubber particles. This kind of barrier can be formed naturally by fatty acids derived from the hydrolysis of fresh latex phospholipids; however to increase stability, additionally commonly used are soaps added at the different stages of latex processing (see 2.3).

2. Processing: from fresh, field latex to concentrated latex

Fresh latex is obtained from a rubber tree by tapping (cutting) a bark, and it is collected into a special cup installed on a tree (**Figure 1**). Making correct incising requires knowledge, precision, and experience. It has to be done carefully and not too deep to avoid damage of a tree. Vessels with latex run spirally in clockwise direction up the tree. Optimal way of tap is from high left- hand to low right hand, which maximizes the number of the opened vessels [5, 11]. These actions have to be carried out early in the morning, when tree "gives" more latex sap due to the greatest turgor of cells during relatively lower transpiration (high humidity and low temperature). On the other hand, in morning conditions, evaporation is lower and latex can stay longer as a stable liquid.

Fresh field latex is a sensitive material, and it coagulates within a few hours after leaving the tree. Fresh latex has different composition than on the next steps of processing (**Table 1**). First of all, it has lower concentration of the rubber material: depend on the clone, it is 30–35% [5]. It has also high variety of non-rubber substances. Characteristics for fresh latex only are, e.g., lutoids. These are vacuolar structures surrounded by a phospholipid membrane containing, among others, significant amounts of proteins [12]. Neutral pH 6–7 and relatively high number of organic compounds, like carbohydrate, make fresh latex sensitive to bacterial growth and putrefaction. Therefore, it is crucial to protect it by the proper preservation.

2.1 Preservation and role of ammonia in latex

The most common preservative for latex is ammonia. Theoretically, addition of 0,2% m/m of ammonia is a minimal dosage, which is good enough for short time protection of collected latex; however, 0,35–0,7% [m/m] is safe regarding the further processing [5]. In practice, fresh latex is sieved to the bulking tanks and then

Figure 1.
Tapped Hevea brasiliensis rubber tree and fresh latex collecting.

Constituents	Fresh latex [%]	Centrifuged latex [%]
Water	59,5–66	38,5–39
Rubber	30–35	min. 60
Non-rubber substances (total)	4–6,5	max. 1,7
Proteinaceous substances	1–1,5	
Resinous substances	1–2	
Carbohydrates	1	
Mineral matter	1	

Table 1.
Composition of fresh natural rubber latex (NRL) and centrifuged latex.

ammonia gas is bubbled through the latex to get a minimum concentration of 1% by weight [13]. The early addition of ammonia is crucial to maintain an optimal property of the latex or rubber material in next stages of processing [14, 15].

Ammonia not only acts as a biocide but also as alkali enhances the negative charge on the rubber particle and improves colloidal stability, works as a buffer, rises the pH, reduces viscosity, and neutralizes free acids formed in the latex.

Additionally, ammonia deactivates some of the metal ions present in the latex (Mg^{2+}, Cu^{2+}), which can have negative effect on the properties and stability in further processing. Copper ions are deactivated by complex formation. Magnesium can be precipitated as magnesium ammonium phosphate. The phosphate comes from the hydrolysis of the phospholipids from natural membrane structures, like rubber particles and lutoids. However, diammonium hydrogen orthophosphate (DAHP) may be added for an excess of magnesium ions over phosphates of natural origin. Phosphates settle down and are left at the bottom of the tank as sludge. It is important to maintain a good magnesium-phosphate balance, as excess phosphate can also bring unexpected difficulties like thickening in further processing.

Because of the odor nuisance at higher concentrations of ammonia, it can be reduced to 0,2% with addition of secondary preservative. The most common and effective is mixture of TMTD and zinc oxide (TMTD/ZnO). Both substances are added to the latex as a well-prepared water dispersion. Small 0,025% dosage of TMTD/ZnO preserves the latex properly, which can be indicated as a lack of volatile fatty acids (VFAs) increase (see 3.9). However, the usage of TMTD requires special care. TMTD is classified as hazardous substance which can be problematic in usage it as a raw material. Additionally, TMTD can be a significant source of limited N-nitrosamines and nitrosatable substances in final dipped products. That is the reason for different secondary preservatives development, such as ZDEC (zinc diethyldithiocarbamate) or TBzTD (tetrabenzyl thiuram disulfide). Latex without TMTD is labeled as "TMTD free" and can be sold finally as a premium latex for safe production (for example medical use).

2.2 Concentration

From the practical and commercial point of view, latex for onward transport to downstream users has to be concentrated. Concentration is a process of increasing the amount of rubber in the latex. According to the accepted standards, the concentrated latex should contain minimum 60% of rubber.

One of the most popular and effective concentrating methods is centrifuging (**Figure 2**). The method uses the difference in density (specific gravity) of the rubber compared to the density of the serum (water phase). The weight of serum (1,02 g/ml) with non-rubber substances is higher than the weight of a rubber (0,92 g/ml). Latex in a centrifugal separator is subjected to high centrifugal force on plates and separated in two fractions: concentrated latex and skim latex. During the process, these two materials go different ways inside the machine to two different outlets. The efficiency of centrifuging process is 85–90% [13], which means skim latex is also a usable material with 2,5–10% of rubber content. Skim fraction can be coagulated and used as a source of dry rubber (skim rubber in blocks or crepe).

Alternative and less popular methods of latex concentration are creaming, evaporation, or electrodecantation. Creaming is a process of separation of rubber and serum based on the difference in weight of these phases, like this occurring during centrifuging. However, during creaming, lighter rubber particles tend to cream up in heavier serum. The velocity of the separation process can be increased by the addition of creaming agent solutions, like sodium alginate or ammonium alginate. Preserved latex mixed well with creaming agents stays undisturbed in vertical creaming tank for approximately 48 hours. After that time, the bottom fraction of skim latex is drained off through the valve, and latex is mixed again. It is worth to mention that this process is used rather for well-ammoniated aged latex (3 weeks after tapping), because the creaming efficiency of fresh latex is lower. Creamed latex, due to the nature of the process it undergoes, has a higher solid content and rubber content but also a higher non-rubber content than centrifuged latex.

Concentrated latex is tested and adjusted to desired specification. Regarding alkalinity (see 3.4), latex can be classified into low ammonia (LA; <0,29% NH_3) and high ammonia (HA; >0,6% NH_3). Currently, the most common commercially used type of LA is LATZ (where "TZ" means TMTD/ZnO preserved). Due to the toxicity of

Figure 2.
Centrifuges used in processing of natural rubber latex.

TMTD, new types of LA, "TMTD free," are developed, but in case of HA with content of ammonia above 0,6%, usage of TMTD can be significantly reduced or eliminated. On the other hand, higher concentration of ammonia can be a problematic issue in the facility of final user, especially in countries with colder climate. Medium ammonia (0,3–0,6% NH_3) is much less popular type of latex.

2.3 Maturation of latex

The period between tapping and packing of the latex is called maturation. In this period of 3–5 weeks, latex enhances its colloidal stability and changes quality parameters. Main and widely used measurement of stability is mechanical stability time (MST) (see 3.5). Mechanical stability of the latex during the first days of processing is relatively low, less than 100 s, which makes it too sensitive for use. Enhancement of MST is an effect of aging, natural processes, and additives used during maturation [16–18]. The natural process of hydrolysis of phospholipids is a source of fatty acids. The soaps they form play an important role in stabilizing the latex. Some amount of soaps is added additionally, which has to be mentioned in protocol from processing. Commonly used ones are ammonium laurate and oleate. Properly adjusted and stored latex increase its MST above 700 s after 3 weeks. The good practice of latex producers is preparing "the MST graph." In this report, producers submit monitored changes of MST result during maturation time with all notes about dosages of additives like soaps or ammonium.

To optimize the maturation process, use of the enzymes is showing good results. The use of lipase, the enzyme responsible for the hydrolysis of lipids and phospholipids, reduces the maturation time for 10–14 days. Additional advantage of enzymatic approach can be the reduction of additives, like soaps, ammonia, or TMTD during processing the latex [18].

3. Natural rubber latex analyses of parameters

The basic specification of concentrated latex is included in norm ISO 2004 [19]. However, in some cases, a more completed characterization is needed. **Table 2** shows all these parameters. The scope of testing and importance of parameters depend on the application of the latex: dipped products, foams, coating, etc.

3.1 Total solid content (TSC)

Total solid content (TSC) of latex is tested by the method described in ISO 124 [20] or ASTM D 1076 [21]. The latex sample is weighed and then dried in an oven (100°C or 70°C) until it completely evaporates, i.e. the mass of sample becomes stable. The result is presented as a percentage base on below equation (Eq. (1)), where m_0 is a weight of latex sample and m_{dry} is a weight of dried sample.

$$TSC[\%] = \frac{m_{dry}}{m_0} \times 100 \tag{1}$$

The result of TSC is needed for subsequent analyses: MST, KOH number, or VFA number. In laboratory practice, a faster but less accurate test can be performed parallel by using a moisture analyzer.

Property	Centrifuged		Creamed		Method
	HA latex	LA latex	HA latex	LA latex	
Total solid content, min. % (by mass)	61,0 or as agreed between the two parties		65,0		ISO 124 [20], ASTM D1076 [21]
Dry rubber content, min. % (by mass)	60,0		64,0		ISO 126 [22]; ASTM D1076 [21]
Non-rubber solids max. % (by mass)	1,7		1,7		—
Alkalinity (as NH_3), calculated with respect to the latex concentrate, % (by mass)	0,60 min.	0,29 max.	0,55 min	0,35 max.	ISO 125 [23]; ASTM D1076 [21]
Mechanical stability, min, seconds	650		650		ISO 35 [24]; ASTM D1076 [21]
Coagulum content, max., % (by mass)	0,03		0,03		ISO 706 [25]; ASTM D1076 [21]
Copper content, max., mg/kg of total solid	8		8		ISO 8053 [26]; ASTM D1278 [27]
Manganese content, max., mg/kg of total solid	8		8		ISO 7780 [28]; ASTM D1278 [27]
Sludge content max., % (by mass)	0,10		0,10		ISO 2005 [29]; ASTM D1076 [21]
Volatile fatty acid (VFA) number, max.	0,06		0,06		ISO 506 [30]; ASTM D1076 [21]
KOH number, max	0,70		0,70		ISO 127 [31]; ASTM D1076 [21]
Parameters out of ISO 2004:					
Magnesium content, max, % (by mass)	0,004 max [32]				ISO 17403 [33], TIS 980–2552 [32]
Phosphate content, max, % (by mass)	—				ISO 19043 [34]
Zinc oxide heat stability (ZHST), min., seconds	300				Test Methods – Revertex Malaysia [35]
Zinc oxide stability (ZST), min., seconds	100				
Zinc oxide viscosity (ZOV_5 and ZOV_{60}) min., %	25				

Table 2.
Requirements of ISO 2004 for centrifuged and creamed natural rubber latex concentrates and additional parameters out of ISO 2004.

3.2 Dry rubber content (DRC)

According to ASTM D 1076 [21] or ISO 126 [22] standards, the latex sample is diluted to 30% TSC and then coagulated with a weak acid (acetic acid 5% or formic acid 2%). After coagulation by the acid, the sample is additionally placed in a hot water bath for 15 minutes. The coagulate is squeezed, washed, and then dried in an oven (70°C or 105°C). Drying the coagulate sample in an oven is time-consuming, so alternatively a quick result can be obtained using microwaves with subsequent verification of the result, using a reference method. Result of dry rubber content (DRC) is presented as percentage base on below equation (Eq. (2)), where m_0 is a mass of liquid latex sample and m_{dry} is a mass of dried coagulum obtained from the sample;

$$DRC[\%] = \frac{m_{dry}}{m_0} \times 100 \quad (2)$$

DRC value is needed to calculate further parameters like VFA number. The DRC value strict refers to the amount of rubber in latex and therefore has an importance from a commercial point of view. Transactions of latex are based on the rubber content, not on the wet material. DRC should be used in calculation of ingredients added into latex as active material quantities (phr) during compounding of the latex (e.g. during prevulcanization).

3.3 Non-rubber solid (NRS)

Non-rubber solid (NRS) is a parameter calculated indirectly as the difference between TSC and DRC. NRS indicates non-rubber components, like minerals or salt content in an aqueous serum phase of latex and residue of chemicals used during processing of field latex. Many non-rubber substances are hydrophilic and may have a negative effect on the conductivity increase of the dipped products, e.g. protective gloves.

3.4 Alkalinity (as NH_3)

The concentration of ammonia is determined by titration of a diluted latex sample with acid (HCl 0,1 N). The method is described in the norm ISO 125 [23]. The result may be expressed as total or as concentration of ammonia in the water phase.

The concentration of ammonia is crucial, both in terms of adequate protection of the latex against microorganisms and in terms of maintaining the colloidal stability of the solution. An irreversible deterioration of latex can be observed when the ammonia concentration drops below 0.2% NH_3. There is an increased risk of coagulation as well as an uncontrolled increase of viscosity during processing. At the same time, the quality, mechanical properties, and the shelf life of the final products are reduced.

3.5 Mechanical stability time (MST)

Mechanical stability time (MST) refers to the resistance of the latex liquid material to the shock caused by the mechanical stress, like mixing or pumping. This stress is imitated by standardized high-speed stirrer (14,000 rpm) of MST machine (**Figure 3**). The test can be done in accordance with the ISO 35 [24] or ASTM 1076. Standard mixers are compatible with both methods.

Sample of latex is diluted to 55% of TSC with ammonia solution (0,6% or 1,6% depend on the alkalinity of the sample) and mixed in MST machine in standard cup until it coagulates. The result is expressed as time (seconds) of mixing the latex till the first visible sign of coagulation occurs. The coagulation process can be observed, for example, by spreading latex on the glass plate during the test. In accordance with the norms, MST should be longer than 650 seconds. On the one hand, too short MST increases the risk of

Figure 3.
MST Klaxon machine.

coagulum forming and thickening during processing of the latex. On the other hand, too long MST (>30 minutes) can be also problematic. Mechanical stability of concentrated latex comes mostly from the soaps naturally formed and soaps added during maturation process (ammonium laurate or potassium oleate). Overstabilization of the latex by the soaps can be the reason of difficulties in dipping process as not effective deaeration and too intensive foaming or less effective gelling on the former.

3.6 Coagulum content

Formation of coagulum in latex is an irreversible process. It is the result of destabilization factors, such as overheating, poor latex stability, and excessive agitation. Method of testing is described in ISO 706 [25] or ASTM D1076. Coagulum content is defined as the percentage mass (on total solids) of the materials, comprising pieces of coagulated rubber, latex skin, and coarse foreign matter retained under the condition of test on a stainless-steel sieve. The latex sample mixed with soap solution (50/50) is poured through a sieve with a mesh size of 180 micrometers (**Figure 4**). The weight of the coagulum is determined from the weight of the sieve dried in the oven.

3.7 Copper and manganese content

Copper and Manganese are prooxidants and accelerate the thermal degradation of raw or vulcanized rubber. Hence, lattices, if contaminated with copper or manganese ions, will have poor aging characteristics.

There are several methods to determinate these metals (**Table 2**) [26–28]. In these methods, dry film needs to be prepared and incinerated. The ash is introduced into solution, and metals are determined spectrophotometrically based on the prepared calibration curve.

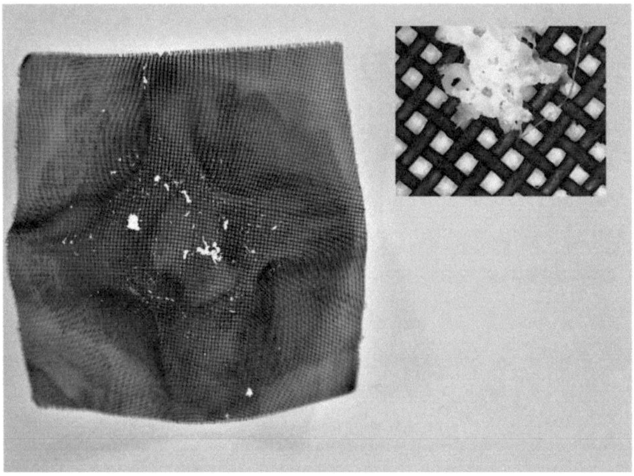

Figure 4.
Sieve with a mesh size of 180 µm after testing the coagulum of latex sample.

3.8 Sludge content

Sludge content refers to the non-polymeric impurities in the latex which tend to sediment under the influence of gravity. In the case of concentrated latex, the sludge is mainly magnesium ammonium phosphate, which remains after the precipitation of magnesium ions with phosphate ions. Magnesium ammonium phosphate content in latex can be kept low by proper desludging of field latex with diammonium hydrogen phosphate (DAHP) before centrifuging (see 2.1).

Test made by ISO 2005 [29] or ASTM D1076 involves centrifugation of the latex followed by repeated washing of the resultant sludge with ammonia-alcohol solution. Finally, residue of the sample is weighted after drying as a mass of the sludge.

3.9 Volatile fatty acid (VFA) number

Volatile fatty acid (VFA) number of latex is defined as the number of grams of potassium hydroxide equivalent to the volatile fatty acid in the latex containing 100 g of total solids.

VFA number is one of the most important parameters for natural latex. It indicates the degree of proper protection of latex against biodegradation. Latex serum contains some carbohydrates (e.g. glucose). Therefore, if latex is not well preserved, bacterial growth occurs. A trace of the presence of microorganisms in latex and their metabolism is an increase in volatile fatty acid concentration—mainly acetic acid but also formic acid and propionic acid. Proper preservation of fresh latex, cleanliness during tapping, and collection are the bases to obtain a low VFA number at subsequent stages.

Method of testing is described in norms ISO 506 [30] and ASTM D1076. The latex is coagulated with ammonium sulfate, and the resultant serum is separated and acidified with sulfuric acid. The serum is steam-distilled on Markham still (**Figure 5**), and the volatile acids (mainly formic acid) present in the latex are determined by acidimetric titration of the distillate. VFA number is calculated from the quantity of barium hydroxide used for titration.

Figure 5.
VFA distillation set with Markham still.

The currently accepted limit for the VFA number according to ISO 2004 is 0,06. However, in previous editions of the standard, with the same method of determination, the limit was higher and was 0,2.

An increase in the concentration of volatile acids goes hand in hand with latex deterioration. The degree of putrefaction can be determined arbitrarily by smelling the latex. The strong smell of ammonia makes it difficult to identify the odor of latex deterioration. The ammonia from the latex sample can be neutralized by a solution of boric acid (5%) before olfactory identification.

3.10 KOH number

The KOH number of latex is defined as the number of grams of potassium hydroxide equivalent to the acid radicals combined with ammonia in latex containing 100 g of total solids. ISO 127 [31] or ASTM D1076 method is based on the titration of the ammonium acid radicals in latex, which has been partially deammoniated with formaldehyde and diluted to 30% TSC. The titration is carried out using 0.5 N KOH solution from about 9.5 to about 12 pH (**Figure 6**). KOH number is calculated based on the inflection point from the titration curve from the equation below (Eq. (3)) where: V is a volume of KOH solution used during titration to obtain inflection point, TSC is a total solid content, M is mole of standard KOH solution and m is a mass of the sample.

$$KOH\ number = \frac{V \times M \times 561}{TSC \times m} \quad (3)$$

In latex, there are present different types of anions. Some of them play a role of stabilizers, like higher fatty acids or protein anions, but the others like volatile fatty

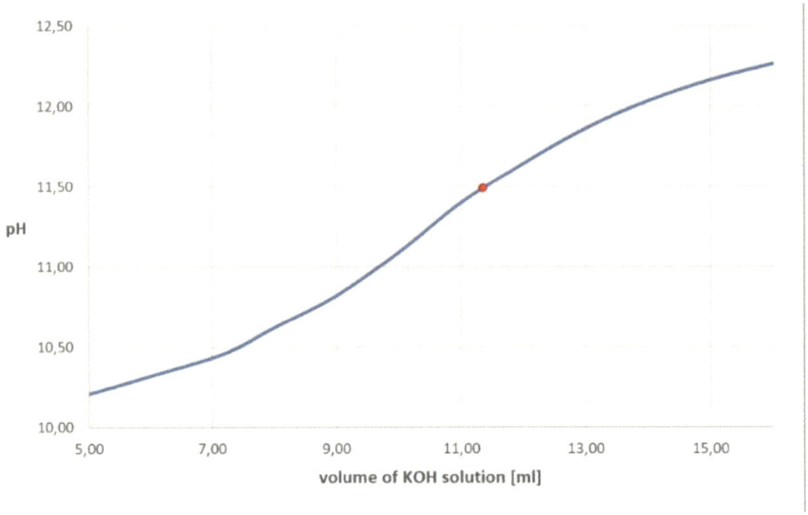

Figure 6.
Example of KOH number titration curve with marked point of inflection.

acids, phosphate, carbonate, citrate, or sulfate can decrease the stability. That is a reason why KOH number result should be related to other parameters as well, to build wider interpretation. Generally, KOH number rises in time, and it can be an indicator of age of latex.

3.11 Viscosity of latex

Viscosity of the latex is presented as kinematic viscosity with unit of millipascals·second (mPa·s) which is equivalent to centipoises (cP) (out of SI system). The most common method of measuring latex viscosity is using a Brookfield viscometer (**Figure 7**). In accordance with ISO 1652 [36] or ASTM D1076, filtered sample of latex, free of air bubbles, is maintained to desired temperature and tested with viscometer. The viscometer is equipped with a set of spindles for measuring different ranges of viscosity. A spindle must be immersed in the tested sample to a designated depth. The apparatus measures viscosity, while the spindle is rotating at a preset speed (rpm). For a more complete comparability of results, additional information about temperature, spindle type, and rotation speed need to be recorded.

Viscosity of latex has great impact on the processability; however, it is not included or specified in ISO 2004. Due to varying customer preferences, limits are usually agreed individually with the latex supplier.

3.12 Magnesium

The presence of magnesium has a markedly negative effect on the quality of latex by contributing to its destabilization. This is due to several issues. Firstly, magnesium reacts with carboxylate ions, forms insoluble soaps with fatty acid, and deactivates their stabilization function. Magnesium also precipitates as magnesium hydroxide. Additionally, it can form primary valence linkages between the surfaces of two adjacent latex particles instead of reacting with two free carboxylate ions which can be a reason of flocculation [37]. In practice, these factors can lead to an uncontrolled

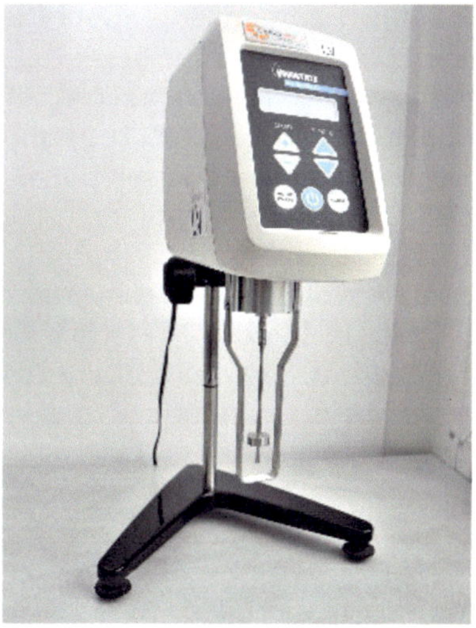

Figure 7.
Brookfield viscometer with spindle number 62. Apparatus in that configuration and rotation speed of 60 rpm has measuring rage 50 – 500 cP.

increase of viscosity or coagulation. As it was mentioned already (see 2.1), to reduce magnesium content, precipitation with phosphates is required (natural or added as DAHP).

Magnesium content is not included in ISO 2004; however, it is a good practice to add the information about the result of free magnesium test in the certificate of analysis for the latex batch. Although there is no international official limit, there are some references. Base on Thai Industrial Standard TIS 980–2552 from 2009, magnesium content should not exceed 40 ppm [32]. This value is similar to the previous literature data where on the studies of Karunanayake L. and Perera G. M. P. [20], 30 ppm of magnesium was acceptable with no significant impact on latex stability. It may be mentioned that a small amount of magnesium can be a desirable evidence of a proper balance being kept when adding phosphates. In other words, a lack of magnesium in the sample may be related to an undesired excessive amount of phosphate used in the earlier stages of latex processing.

According to ISO 17403 [33], the determination of magnesium is performed by complexometric titration with EDTA. In case of concentrated latex, the determination is carried out in a serum sample (after coagulation). Field latex may be analyzed without separation (coagulation is not needed).

Additionally, to improve the correct magnesium-phosphate balance in field latex, quick test kits may be useful. In case of this solution, result of magnesium content can be obtained within a few minutes [38].

3.13 Phosphates in latex

The phosphate issue has been already described in previous parts (see 2.1; 3.12). On the one hand, it is largely linked with the reduction of magnesium; but on the other hand, too high excess of phosphates can have a negative impact on latex stability and cause problems of thickening during further processing [38]. As there is no officially defined limit value for phosphate content, the best approach is to determine it individually for the particular process and latex application. Some manufacturers of dipped products in Thailand consider that the best properties of the latex film are achieved when the amount of phosphates do not exceed 100 ppm [39]. However, in the study of Karunanayake L. and Perera G. M. P., the sample containing small excess of 30 ppm phosphate relative to magnesium showed the highest mechanical stability and best properties after prevulcanization. Moreover, further samples with higher amounts of phosphate showed increasingly poorer stability.

The conventional determination of phosphates in latex is carried out using the spectrophotometric method described in ISO 19043 [34]. The latex sample is coagulated to allow separation of the serum. Sample of filtered serum with developed color is tested on spectrophotometer. The determination is carried out based on a previously prepared calibration curve.

3.14 Zinc oxide stability test (ZOV, ZST & ZHST)

Chemical stress caused during processing by the addition of compounding mixtures should not be a reason for the significant destabilization of latex. Zinc oxide stability tests are a group of tests which are indicators of the latex resistance to this kind of stress. The method is not standardized in ISO but can be helpful as an additional indicator during latex quality control. The basis of the method is an excessive addition of zinc oxide water dispersion (ZnO 40%) in excess, which cause changes in the latex properties (1 phr). Destabilized sample is tested for heat resistance and for changes in viscosity and mechanical stability. Practically, one sample with ZnO (40%) addition is prepared and finally split for all the tests.

Zinc Oxide Heat Stability Test (ZHST) theoretically is a measure of latex resistance to heat. The heat can be an issue, for example, during warm compounding and warm prevulcanization (more details on point 4). In this test, 1 hour after 1 phr ZnO addition to the latex, the sample of 50 g is stirred in warm condition of water bath at 90°C. Result of a test is presented as a time of stirring till final coagulation occurs. Recommended ZHST should exceed 300 seconds [35].

Zinc Oxide Stability Test (ZST) can be some indication of proper resistance of the latex to decrease mechanical stability after compounding and during further processing. Latex sample, 1 hour after 1 phr ZnO addition, is tested on MST (see 3.5). Result of the test is the time of mixing with MST machine till final coalescence (coagulation). Recommended ZST should exceed 100 seconds [35].

Zinc Oxide Viscosity (ZOV) is a measure of latex resistance to drastic changes in viscosity after compounding and during further processing. In case of ZOV, addition of ZnO dispersion changes the viscosity immediately. The measurements of ZOV are percent changes of Brookfield viscosity 5 min (ZOV_5) and 1 hour after 1 phr ZnO addition (ZOV_{60}). It is recommended that increase on viscosity should not exceed 25% [35].

4. Compounding and prevulcanization of NRL

Latex in the non-vulcanized state is not suitable for manufacturing latex goods, as it does not have required flexibility, resistance, and tensile properties. Those properties are obtained during vulcanization process. The term and principle of this process were introduced by Charles Goodyer, who discovered that using sulfur and temperature changes properties of natural rubber. The term itself comes from the Roman god of fire—Vulcan [40]. In a simple way, vulcanization means creation of crosslinking between free chains of rubber into network, by adding special composition of chemicals and heat. Addition of chemicals into the latex is called compounding [11]. This is one of the most important processes in the preparation of latex to be ready for use. Quality of used chemicals, which are used in a form of water-based dispersions, as well as the proper handling, like agitation, time, and temperature, have big impact on the final properties of ready-for-use latex. The process and composition of vulcanizing agent are strictly confidential know-how of each company. Latex after compounding can be used immediately, or can be left for a period, that the crosslinks are created. The first case requires post-vulcanization with use of high temperature. The second case is called prevulcanization and can be conducted as hot prevulcanization in elevated temperature (up to 80°C) for couple of hours or cold prevulcanization in ambient temperature for couple of days. Lower prevulcanization temperatures give generally better physical properties of finished products [11]; however, in the end, the choice of system depends on many factors and is compromised between ability for storage capacity, timing, and heating costs during prevulcanization, during curing and desired final properties.

Most commonly used vulcanization system is sulfur vulcanization, with the use of accelerators from group of dithiocarbamates (zinc diethyldithiocarbamate, zinc dibenzyldithiocarbamate, and zinc diisononyldithiocarbamate), thiurams (tetrabenzylthiuram disulphide and tetradimethylthuiram disulphide), thiazoles (zinc 2-mercamtobenzothiazole), and zinc oxide as activator. Additionally, stabilization system is required to avoid losing of colloidal stability of compounded latex. For this reason, KOH is widely used, or specific stabilizers commercially offered by specialized companies. Use of specific prevulcanization systems ensures that final product obtains required mechanical and physical properties, as well as functional properties.

Mechanism of crosslinking formation is not fully known yet; nevertheless, the basic of this complexed reaction can be explained in several steps. First step is the creation of active sulfurating agent from sulfur, accelerator, and zinc oxide. Next step is the reaction of sulfurating agent with the rubber molecule and creating rubber-bound intermediates, which react further with other rubber molecules, that is, creation of initial crosslinking network. In further stage, the network matures, as sulfur crosslinks exchange. In the end, matured crosslink network is formed [11, 40].

It is worth to mention that most of the accelerators are precursors of N-nitrosamines, which can be cancerogenic as presented above specific limit values, depending also on the exposure scenario for certain products. The industry must follow up with recent research studies and legislation changes and adjust the technology accordingly [41–44].

In latex compounding, the quantity of all ingredients is added in units called phr—parts per hundred parts of rubber. This gives clear indication on amount of added active substance into latex of known rubber content.

4.1 Monitoring of prevulcanization process

Prevulcanization process is a kind of chemical stress for latex. That might influence the colloidal properties of latex, as increase of viscosity and decrease of mechanical stability time and appearance of coagulum. It is important to monitor the latex properties during prevulcanization, as well as the prevulcanization stage [11, 13, 45].

There are several methods for the determination of crosslinking formation. Most popular ones are chloroform number, toluene swell, and modulus as part of tensile testing.

4.1.1 Chloroform number

Chloroform number is the easiest and fastest testing method. Equal amount of latex and chloroform are mixed, and during stirring the latex coagulates. Created coagulum has different appearance, from soft and tacky mass to almost dry, crumbled particles, depending on the prevulcanization stage (**Figure 8**). The chloroform number is described as number from 1 to 4:

1. soft, tacky, and very stretchy mass

2. soft and bit tacky mass, but breaks when stretched

3. not tacky clump, not very stretchy, easily breaks pulled apart

4. dry, crumbled particles

The weak point of this method is subjectiveness, and not exact and standardized numbers. The performers often use additionally half numbers, for example 2,5 or 3,5 [11, 46].

4.1.2 Swelling test

Swelling test, with use of toluene as most suitable solvent, is another fast-testing method for determination of crosslink density. Small quantity of latex is poured on glass plate and with the use of steel bar spreaded equally. After drying, the sheet of latex prior dusting is removed from the plate, and with use of die cutter a circle sample is cut. This sample is immersed in toluene and absorbs solvent. The less the crosslinks in the latex, the more solvent the sample can swell. After certain time, usually 20–45 minutes, diameter or weight of swollen sample is measured and compared to non-swelled sample. Stage of prevulcanization can be determined based on equations below (Eqs. (4, 5)), where Q is a swell ratio calculated from the change of the sample weight (W1 – initial weight; W2 – swollen weight) and L is a swell ratio calculated from the change of the sample diameter (l1 – initial diameter; l2 – swollen diameter) [46]:

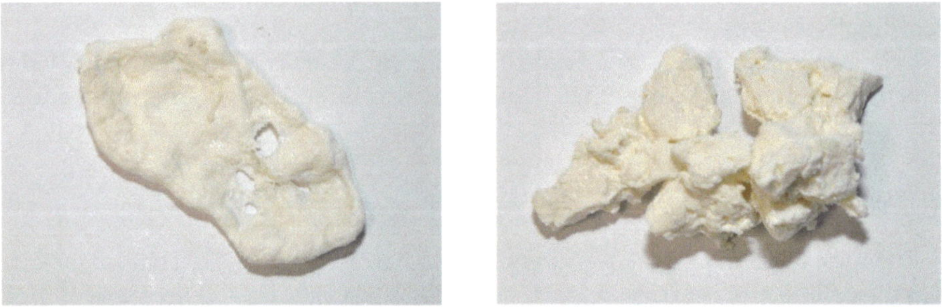

a. b.

Figure 8.
Sample of unvulcanized LATZ (a.) and vulcanized PV (b.) treated with chloroform during chloroform number testing.

$$Q = \frac{W_2 - W_1}{W_1} \tag{4}$$

$$L = \frac{l_2}{l_1} \tag{5}$$

The interpretation of the numerical value of the calculated ratio is shown in the **Table 3**.

4.1.3 Modulus

Modulus is the most accurate parameter, however requires specific instrument, which is a tensile machine. It can be conducted as extensive test of relaxed modulus or tensile modulus. For testing relaxed modulus, latex ring is formed by dipping a tube in latex and then, after rapid drying, modulus at 100% extension is checked. Tensile modulus requires forming latex film and then cutting the sample with dumbbell cutter and checking the modulus at 300%, 500% or 700% extension [11, 46].

4.2 PV grades based on tensile properties

Different applications require latex of different prevulcanization stage or level. The prevulcanized latex is offered also commercially by specialized producers. While during manufacturing of latex goods, the producers use PV specified with chloroform number, the commercially offered prevulcanized latexes are specified or described with modulus (modulus at 700%) tested on tensile machine (**Figure 9**) [47]. The commercially available grades of PV are as follows:

Low modulus (8,0–10,5 MPa)—suitable for latex balloons, easy to inflate.
Medium modulus (11,0–13,5 MPa)—suitable for latex gloves.
High modulus (14,0–17,5 MPa)—suitable for condoms and catheters.

5. Dipping process as production technology of latex goods

Dipping technology is one of the most popular methods of producing thin-walled latex products. Products such as latex gloves, balloons, condoms, and catheters are produced with this technology. The process might be conducted as batch dipping process or chain (continuous) dipping process, with the use of formers made of different

Stage of prevulcanization	Q	L
Unvulcanized	>15	>2,6
Lightly Vulcanized	7-15	2,0-2,6
Moderately Vulcanized	5-7	1,8-2,0
Fully Vulcanized	<5	<1,75

Table 3.
Interpretation of the swelling index results.

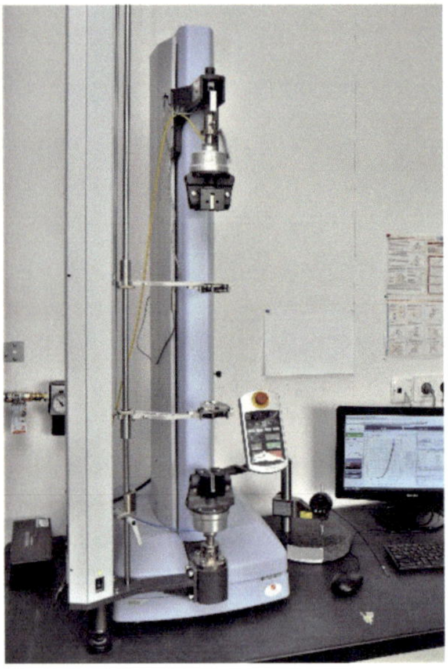

Figure 9.
Tensile machine during test of rubber sample.

materials—they can be ceramic, plastic, metal, or glass. Batch dipping is mainly used for irregular shapes or thicker products, like industrial or household gloves and gives lower productivity, while chain dipping is used for mass production volumes of less demanding shapes [13]. Applying latex to the former can be done with different ways, from which the most popular are coagulant dipping (gloves and balloons) and straight dipping (condoms) [48, 49].

In coagulant dipping, at the first stage, the cleaned formers are dipped in coagulant solution, consisting mainly of water, calcium nitrate, and powder, which can be calcium carbonate, talc, and diatomaceous earth. Coagulant creates thin layer on the former, which is then dipped in liquid latex. Calcium ions from coagulant influence liquid latex colloidal stability, causing gelling of latex, which forms thin latex film covering the former. Powder in coagulant prevents from sticking of the film to the former and in the end enables removal of ready products from the formers in the last production stage.

Next step is leaching of latex film with water. In this step, residue of chemicals from latex and coagulant are leached out, as well as proteins, naturally occurring in latex. This step is called also wet gel leaching.

After leaching in water, products are dipped in anti-tack agent, which prevents from sticking outside. For this purpose, composites of soaps or dispersions of powders are used. Then the edges of dipped film are formed by rolling brushes, creating beads. Bead enables good grip and inflation of a balloon or good donning of a glove.

Finally, formers with latex film go for about 1 hour into drying oven, where temperature gradually increases to approx. 90°C. Latex film needs to be finally cured and dried, as it contains about 60% of water. After passing the oven, latex products are stripped, manually or automatically, with the use of compressed air. Water stripping is also used by some producers. After this stage, balloons or gloves are still slightly humid and require final drying in the tumble dryers [50]. The scheme of dipping process flow is presented in **Figure 10**.

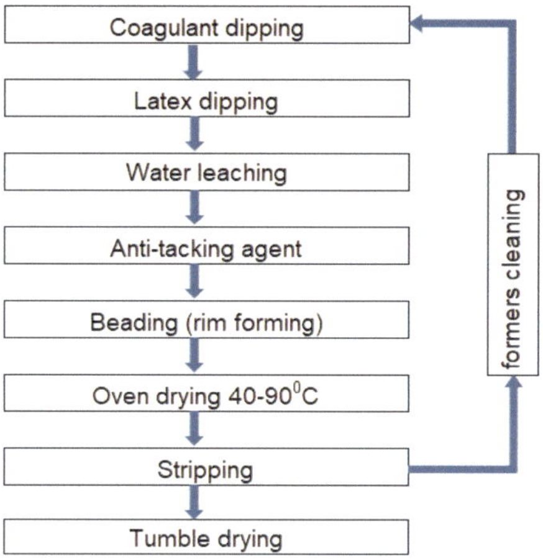

Figure 10.
The scheme of dipping process flow.

6. Trends and challenges in terms of sustainability and environment

Couple of years ago, the main challenge for the industry manufacturing latex goods was that they were required to be resistant to aging and had very long shelf life. In recent times, the trends are going into another direction. Products and the manufacturing process aim to have reduced impact on the environment.

6.1 Classification of modified latex–Single Use Plastics Directive

The production and distribution of latex balloons, as well as other natural latex products, can be limited in the coming years due to environmental impact requirements for latex waste. Latex balloons have already been included in the scope of Directive (EU) 2019/904 of the European Parliament and of the Council on reducing the environmental impact of certain plastic products [51]. Market trends are moving toward a pro-environmental direction and the action of the European Union and individual Member States toward tightening regulations and increasing producer responsibility. Additionally, environmentally friendly products are more often chosen by consumers and end users [50].

6.2 Carbon footprint

In the view of the challenge mentioned in point 6.1, it is worth to highlight the natural plant origin of latex, as many of the end users are not aware of it. Natural rubber latex polymerization occurs in trees, which is different from the synthetic polymerization. The rubber trees during their life cycle absorb carbon dioxide from the environment. On the other hand, cultivation of rubber trees on the plantations and further processing of field latex create emission of carbon dioxide. This data can be used for the calculation of carbon footprint of latex or latex products, which is total emission of greenhouse gas (GHG) emissions expressed as equivalent of carbon dioxide CO_2.

Latex producers and suppliers are aiming for carbon neutral, often with successful outcome. Concerning goods produced from latex, studies on the carbon footprint are conducted, focusing on carbon dioxide emission during processing and manufacturing, following Life Cycle Assessment (LCA) and Green Houses Gases Protocol (GHG Protocol) [52–54].

6.3 Sustainable sources

Global demand for natural rubber created expansion of rubber plantations, which has been a driver of deforestation in the past. Currently, there are organizations and standards, which encourage the producers to proceed in responsible way. One of them is Forest Stewardship Council—FSC. The core of FSC certification is to ensure and confirm that products are deforestation-free and socially responsible, and that there is full transparency and traceability from the plantation to the final product and end user [55]. In 2017, 4% of total rubber plantation areas were FSC-certified [56].

Other recognizable association that focuses on social part of sustainable approach for the improvement of the working and living conditions of the primary producers of natural latex is the Fair Rubber Association. Similar to FSC, products like mattresses, condoms, balloons, and rubber boots are also certified according to the designated standards [57].

7. Conclusion

The processing of latex due to its natural origin is an extremely challenging issue. Users in the whole supply chain need to consider many factors in a holistic view, from the early stages of latex production on the plantation till the final use. Understanding the latex properties and their influence on further processing is important for obtaining stable process and good quality of final product.

Despite of demanding nature of this raw material, requiring experience in handling, it is worth to gain an insight into latex application. Plant origin of natural rubber latex gives an opportunity for sustainable approach and solutions described in the chapter.

Author details

Jacek Rafał Kędzia*, Anna Maria Sitko, Józef Tadeusz Haponiuk and Justyna Kucińska Lipka
Gdansk University of Technology, Poland

*Address all correspondence to: jacek.kedzia@pg.edu.pl

References

[1] Ciechanowicz L. The world of elastomers the Chronicle of Events from 1495 to 2000. Elastomery. 2001;**5**(2):3-10

[2] Rzymski WM. Kauczuk naturalny: nieco historii, stan obecny i perspektywy. Wybrane zagadnienia. Elastomery. 2013;**17**(3):12-20

[3] Sowcharoensuk C. Natural Rubber Processing. Thailand Industry Outlook. Apr 2021. Available from: https://www.krungsri.com/en/research/industry/industry-outlook/agriculture/rubber/IO/io-rubber-21 [Accessed 01 Jun 2022]

[4] Wei Y, Zhu D, Xie W, Xia J, He M, Liao S. In-situ observation of spatial organization of natural rubber latex particles and exploring the relationship between particle size and mechanical properties of natural rubber. Industrial Crops and Products. 2022;**180**:114737

[5] Blackley DC. Polymer Latices science and technology. In: Types of Latices. 2nd ed. Vol. 2. Dordrecht: Springer Science + Business Media; 1997

[6] Nawamawat K, Sakdapipanich JT, Ho ChC, Ma Y, Song J, Vancsoet JG. Surface nanostructure of Hevea brasiliensis natural rubber latex particles. Colloids and Surfaces A: Physicochemistry Engineering Aspects. 2011;**390**:157-116

[7] Sakdapipanich JT. Structural characterization of natural rubber based on recent evidence from selective enzymatic treatments. Journal of Bioscience and Bioengineering. 2006;**103**(4):287-292

[8] Tarachiwin L, Sakdapipanich J, Ute K, Kitayama T, Bamba T, Fukusaki E, et al. Structural characterization of alpha-terminal group of natural rubber. 1. Decomposition of branch-points by lipase and phosphatase treatments. Biomacromolecules. 2005;**6**(4):1851-1857

[9] Sriring M, Nimpaiboon A, Kumarn S, Sirisinha C, Sakdapipanich J, Toki S. Viscoelastic and mechanical properties of large- and small-particle natural rubber before and after vulcanization. Polymer Testing. 2018;**70**:127-134

[10] Singh M. The colloidal properties of commercial natural rubber Latex concentrates. Journal of Rubber Research. 2018;**21**:2

[11] Hill DM. The Science and Technology of Latex Dipping. UK: Smithers Rapra Technology; 2018

[12] Subroto T, Vries H, Schuringa JJ, Soedjanaatmadja UMS, Hofsteenge J, Jekel PA, et al. Enzymic and structural studies on processed proteins from the vacuolar (lutoid-body) fraction of latex of Hevea brasiliensis. Plant Physiology and Biochemistry. 2001;**39**:1047-1055

[13] Joseph R. Practical guide to latex technology. UK: Smithers Rapra Technology; 2013

[14] Santipanusopon S, Riyajan S. Effect of field natural rubber latex with different ammonia contents and storage period on physical properties of latex concentrate, stability of skim latex and dipped film. Physics Procedia. 2009;**2**:127-134

[15] Salomez M, Subileau M, Intapun J, Bonfils F, Sainte-Beuve J, Vaysse L, et al. Micro-organisms in latex and natural rubber coagula of Hevea brasiliensis and their impact on rubber composition, structure and properties. Journal of

Applied Microbiology. 2014;**117**:921-929. DOI: 10.1111/jam.12556

[16] Yu H, Wang Q, Li J, Liu Y, He D, Gao X, et al. Effect of lipids on the stability of natural rubber latex and tensile properties of its films. Journal of Rubber Research. 2017;**20**:213-222. DOI: 10.1007/BF03449153

[17] Pendle TD. Production, properties and stability of NR latices. Rubber Chemistry and Technology. 1989;**63**:234-243

[18] Wongkhat J, Pattamaprom C. The potential of enzyme in reducing the maturation time and the use of ammonia in concentrated natural rubber Latex. IOP Conference Series: Materials Science and Engineering. 2019;**652**:012040. DOI: 10.1088/1757-899X/652/1/012040

[19] ISO 2004:2017 – Natural rubber latex concentrate - Centrifuged or creamed, ammonia-preserved types – Specifications. Available from: https://cdn.standards.iteh.ai/samples/70281/23b5bedd95e1478c978169e9039b7eb2/ISO-2004-2017.pdf

[20] ISO 124:2014. Latex, rubber — Determination of total solids content

[21] ASTM D1076-21. Standard Specification for Rubber—Concentrated, Ammonia Stabilized, Creamed, and Centrifuged Natural Latex

[22] ISO 126:2005. Natural rubber latex concentrate — Determination of dry rubber content

[23] ISO 125:2020. Natural rubber latex concentrate — Determination of alkalinity

[24] ISO 35:2004. Natural rubber latex concentrate — Determination of mechanical stability

[25] ISO 706:2004. Rubber latex — Determination of coagulum content (sieve residue)

[26] ISO 8053:1995. Rubber and latex — Determination of copper content — Photometric method

[27] ASTM D1278-91a. 2020. Standard Test Methods for Rubber from Natural Sources—Chemical Analysis

[28] ISO 7780:1998. Rubbers and rubber latices — Determination of manganese content — Sodium periodate photometric methods

[29] ISO 2005:2014. Rubber latex, natural, concentrate — Determination of sludge content

[30] ISO 506:2020. Rubber latex, natural, concentrate — Determination of volatile fatty acid number

[31] ISO 127:2018. Rubber, natural latex concentrate — Determination of KOH number

[32] TIS 980-2552. Thai Industrial Standard-Natural Rubber Latex Concentrate. Bangkok: Thai Industrial Standards Institute; 2009

[33] ISO 17403:2014. Rubber — Determination of magnesium content of field and concentrated natural rubber latices by titration (cyanide-free method)

[34] ISO/DIS 19043. Natural rubber latex concentrate — Determination of total phosphate content by spectrophotometric method

[35] Test Methods – Revertex Malaysia Sdn Bhd, Kluang, Malaysia

[36] ISO 1652:2011. Rubber Latex— Determination of Apparent Viscosity by the Brookfield Test Method

[37] Karunanayake L, Perera GMP. Effect of magnesium and phosphate ions on the stability of concentrated natural rubber Latex and the properties of natural rubber Latex–dipped products. Journal of Applied Polymer Science. 2006;**99**:3120-3124. DOI: 10.1002/app.22944

[38] Malahom N, Jarujamrus P, Meelapsom R, Siripinyanond A, Amatatongchai M, Chairam S. Simple test kit based on colorimetry for quantification of magnesium content in natural rubber latex by miniaturized complexometric titration without using masking agent. Polymer Testing. 2017;**59**:160-167

[39] Sirikarn J. Method Development for Determination of Phosphate in Natural Rubber [thesis]. Thailand: Prince of Songkla University, Analytical Chemistry; 2012

[40] Joseph AM, George B, Mdhusoodanan KN, Alex R. Current status of Sulphur vulcanisation and devulcanization chemistry - process of vulcanization. Rubber Science. 2015;**28**:82-121

[41] Pinprayoon O, Mae W. Migration of N-nitrosamines from rubber gloves for handling food - effect of extraction media. In: Proceedings of the International Rubber Conference (IRC'18) IOP Conf. Series: Materials Science and Engineering 548 (2019) 4-6 September 2018. Kuala Lumpur, Malaysia: IOP Publishing; 2019. DOI: 10.1088/1757-899X/548/1/012022

[42] BgVV. Risk assessment of N-nitrosamines in balloons [Internet]. 2002. Available from: http://www.bfr.bund.de [Accessed 18 Jun 2022]

[43] VWA/KvW Report nr. ND04o063/02 Migration of N-nitrosamines and N-nitrosatable substances from latex balloons [Internet]. 2005. Available from: http://www.nwa.nl [Accessed 18 Jun 2022]

[44] SCCP. Opinion on the Presence and Release of Nitrosamine and Nitrosatable Compounds from Rubber Balloons [Internet]. 2007. Available from: https://ec.europa.eu/health/ph_risk/committees/04_sccp/docs/sccp_o_121.pdf [Accessed 18 Jun 2022]

[45] Sasidharan KK, Rani JR, Palaty S, Gopalakrishnan KS, Rajamma G, Viswanatha Pillai P. Effect of the vulcanization time and storage on the stability and physical properties of sulfur-Prevulcanized natural rubber Latex. Journal of Applied Polymer Science. 2005;**97**:1804-1811. DOI: 10.1002/app.21918

[46] Gorton ADT, Pendle TD. A new rapid measurement of crosslink density in compounded natural latices. NR Technology. 1976;**7**(Part 4):77-81

[47] PV for various application [Internet]. Available from: http://www.getahindus.com.my [Accessed 18 Jun 2022]

[48] Sasidharan KK, Rani Joseph R, Rajammal G, Viswanatha Pillai P, Gopalakrishnan KS. Studies on the dipping characteristics of RVNRL and NR latex compounds. Journal of Applied Polymer Science. 2001;**81**(13):3141-3148. DOI: 10.1002/app.1766

[49] Chirinos HD, Guedes SML. The manufacture of gloves using RVNRL: Parameters of the coagulant dipping process. Brazilian Journal of Chemical Engineering. 1998;**15**(4). DOI: 10.1590/S0104-66321998000400003

[50] Sitko A, Sienkiewicz M, Kucińska-Lipka J, Kędzia J, Hrycyk K. Natural rubber latex balloons - biobased product

struggling environmental policies and customer choices. In: Proceedings of the International Scientific-Technical Conference (APGIP-11); 16-20 May 2022; Ukraine. Lviv; 2022. pp. 238-240

[51] Directive (EU) 2019/904 of the European Parliament and of the Council of 5 June 2019 on the reduction of the impact of certain plastic products on the environment. Available from: https://www.legislation.gov.uk/eudr/2019/904/introduction

[52] Usubharatana P, Phungrassami H. Carbon footprints of rubber products supply chains (fresh latex to rubber glove). Applied Ecology and Environmental Research. 2018;**16**:1639-1657. DOI: 10.15666/aeer/1602_16391657

[53] Maik BM, Lehmann AK, Naranjo E, Finkbeiner MA. Condom's footprint - life cycle assessment of a natural rubber condom. International Journal of Life Cycle Assessment. 2020;**25**:964-979. DOI: 10.1007/s11367-019-01701-y

[54] Greenhouse Gas Protocol [Internet]. Available from: http://www. ghgprotocol.org

[55] Forest Stewardship Council [Internet]. Available from: http://www.fsc.org [Accessed 17 Jun 2022]

[56] FSC®-certified natural rubber: Deforestation free, socially responsible [Internet]. 2017. Available from: https://ic.fsc.org/file-download.fscr-certified-natural-rubber-deforestation-free-socially-responsible.a-2963.pdf [Accessed 17 Jun 2022]

[57] About the fair rubber association [Internet]. Available from: http://www.fairrubber.org [Accessed 17 Jun 2022]

Chapter 6

Characterization of Hydrogenated Amorphous Silicon Using Infrared Spectroscopy and Ellipsometry Measurements

Mounir Kassmi

Abstract

We described the primary mixed compositions of hydrogenated amorphous silicon on the surface of glass (7059) in this chapter and distinguished them optically by combining the outcomes of infrared spectroscopy and ellipsometric tests. The particular hydrogen content of the aspherical voids created determines the energy level of the optical band, which ranges from 1 eV to 4 eV depending on how passivated or unpassivated the composition is. Additionally, the dielectric response is influenced by the size and proportion of the vacuum occupation relative to the surrounding phase, and each dielectric response is based on how much the implicated components have been passivated.

Keywords: hydrogenated amorphous silicon, infraredspectroscopy, ellipsometry, band gap energy, void, passivation, vibrational mode

1. Introduction

Researchers and professionals have focused a lot of attention on photovoltaic power since it was first introduced in the engineering area in order to create materials with practical qualities for the conversion of solar-powered energy. For a variety of reasons, including its consistency in quantity, the ease with which it may be elaborated, and its safety, silicon emerged as the most exciting newcomer in the universe of these materials with relation to this transformation cycle [1]. This material has been researched in a number of different forms, including mono-crystalline silicon, protocrystalline silicon, hydrogenated polymorphous silicon, and hydrogenated amorphous silicon [2]. A few endeavors used infrared spectroscopy and ellipsometry measurements to control the optical characteristics of this intriguing material [3]. In addition, it is imperative that numerous studies be conducted in order to accurately link the hydrogen concentration and the presence of microvoids within this material, using both ellipsometry measurement and other methods [3–7]. The latter has, however, clearly superior characteristics to the others, including a large absorption coefficient and a direct band gap energy that are easily adjustable using a variety of elaboration approaches by adjusting the temperature and hydrogen flow [7–9]. A lot of research on the a-Si: H material has been published in the literature over the past three decades, and it has produced a number of intriguing outcomes that have led to the conclusion that this material has good absorption properties. Besides, it has poor transport properties with a short carrier diffusion length of around 300 nm [10] and there is 10–30% efficiency degradation under light soaking owing to the Staebler-Wronski effect [11, 12].

The desired quality for the proper functioning of solar cells based on thin layers of hydrogenated amorphous silicon still presents a challenge for researchers despite the efforts made and the results obtained during this time because there are still a number of phenomena that are not fully understood, such as the impact of the Si—H bond on the gap energy, microvoid density, optical index, and transport properties [5, 13]. A small-angle X-ray scattering technique (SAXS) provides details of the void covered in the array [14]. It was determined that the optoelectronic characteristics on the surface of glass are particularly influenced by the size and density of voids, and that voids reduce the mass density of such materials [15]. We are better able to comprehend the root of the enhanced transport qualities to the thorough and precise description of the film structure. Many earlier studies have established the importance of hydrogenation in the enhancement of specific properties, as hydrogen decreases defects by reducing the number of dangling bonds, which are responsible for subpar device performance [16, 17]. The physical understanding of this phenomenon is still a long way from being in a place where it can be controlled in order to optimize thin layers. The band gap may be another topic of contention in the current state of science between researchers, some of whom concur with a monotonic decreasing relationship for structural disorder [18, 19] and a monotonic increasing relationship for hydrogen concentration [20, 21]. As a result, some scientists came to the conclusion that neither the hydrogen concentration nor the structure disorder has a standalone effect on the gap.

The interdependency of the two factors is what makes these conversations challenging [22, 23]. Other earlier studies have demonstrated that as the temperature of the thin layer deposit rises, the gap and overall structural material disorder are reduced significantly [24, 25]. The hydrogen concentration and the structural disorder are both influenced by the substrate's temperature, and both variables are required to explain an observed local minimum of gap, but the silicon monohydride (Si—H) bond density only accounts for this dependency [26]. All of these outcomes support the assertion that even though research on the a-Si: H material has improved in terms of the latter's ability to be doped to increase its transport capabilities [27, 28], it's still challenging to regulate the factors that affected its optical qualities. Instead, because much less material is needed to respond and totally absorb the light, ultra-thin film optoelectronic devices, particularly those built of hydrogenated amorphous silicon a-Si: H, have the potential to be less expensive. The existence of voids and hydrogen bonding are further topics for discussion. Smets et al [4] have shown that when the amount of hydrogen connected to silicon exceeds 14%, the material may contain microvoids, while less than that, it mainly contains vacancies decorated by hydrogen.

Thus, we may determine the film microstructure and the likelihood of occurrence of such configuration, whether isolated or related to another, using the mass density of the film and the intensity of infrared absorption modes. A significant portion of earlier work relied on a trial-and-error method of hydrogenating amorphous silicon, such as using low hydrogenation gas esteems ranging from 2 to 75 sccm, which produced a variety of results. Among these are the characteristic farthest reaches of the existence or absence of vacancies, which are associated with a critical hydrogen concentration (14%), as well as the relationship between the bandgap energy and the density of Si—H bonds [26].

In amorphous silicon, streams of hydrogenation gas that did not exceed 75 sccm were previously operated. In this chapter, a significant amount of hydrogenation gas —roughly 200 sccm—is sufficient to determine the film's properties in terms of both its fundamental structure and its optoelectronic properties. Besides, we endeavor to follow, from one perspective, the connection between the measures of hydrogen related to the dimensions of the round voids framed in the film folds, and similarly, the relationship that influences the difference in bond density with the components of the various shapes, considering the infrared vibration frequencies of the bonds. At room temperature, ellipsometric measurements and infrared spectroscopy were used to account for the network's hydride arrangements. For all tetrahedral configurations, experimental data were examined to clarify the spectral dependency of dielectric functions. The goal of this study is to learn more about the many ways silicon and hydrogen atoms attach using observed optical functions in a-Si: H. This research uses the well-known Bruggeman model (EMA), which has been applied to composite and heterogeneous medium. This research does suggest that optical constant measurements may be a very sensitive probe of microscopic compositions and by the very compact compound matrix within the different heterogeneous formations.

2. Methodologies for analysis and synthesis

Ultrathin a-Si: H films were produced using RF-PECVD. Silane (4 sccm) and hydrogen were used as gas sources for the deposition (gas mixture). A common deposition parameter uses a high radio frequency power value and a 750 mTorr starting pressure (60 W). The hydrogenation gas flow rates were 100 and 200 sccm. The substrate temperature was maintained at 300 °C for the whole 30-minute deposition period.

Using the CuK_α line ($\lambda = 1.54056\mathring{A}$), X-ray diffraction analysis (Bruker D8 Advance, Germany) was used to examine the amorphous structure of hydrogenated silicon. Hydrogen concentration was determined using IR absorption tests using a Perkin-Elmer FTIR in the 400–4000 cm^{-1}. Real and imaginary components of dielectric functions were determined using ellipsometry measurements in the spectral range [1–5 eV].

In order to increase the measurement's sensitivity, an incidence angle θ_0 is chosen. However, depending on the optical constants of the samples, the incidence angle is changed. As a result, a value of 70° for θ_0 may allow us to reach the maximum amount of reflected light with the sample and also produce a significant oscillation in the Cos (Δ) spectrum. Data collection and analysis were carried out utilizing WinElly II software (version 2.0.0.0), which presumes a completely flat surface.

3. X-ray diffraction analysis

The effects of hydrogen on the network have been extensively studied in the literature even though this topic has previously been brought up thirty years ago [29, 30]. The X-ray data revealed an amorphous structure along with a strong impact

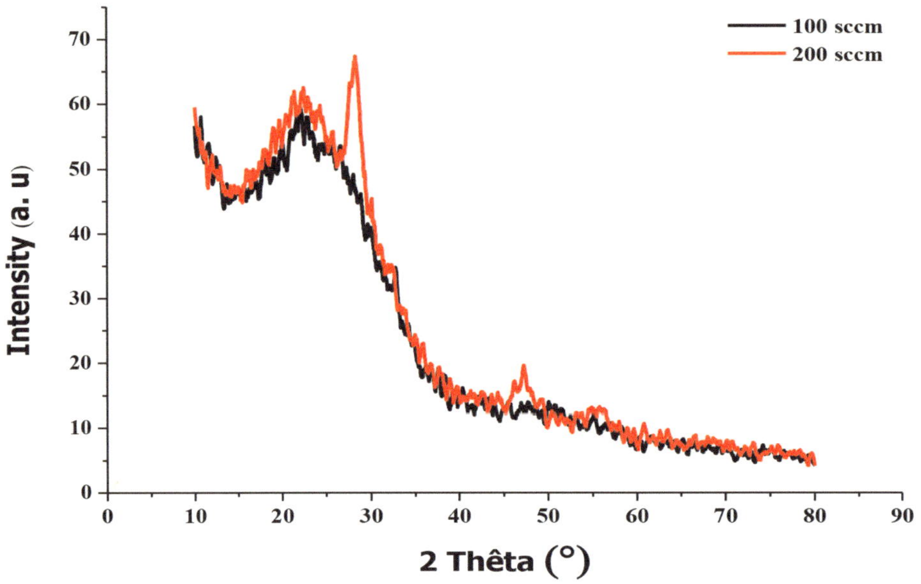

Figure 1.
X-ray diffraction patterns of a-Si: H thin films.

of hydrogen on the structure as seen by the numerous updated Si—H bonds [31]. The presence of a metastable network with various optical characteristics in the film is indicated by this. Additionally, compared to tetrahedral configurations without hydrogen, those that are distinguished by their thickness are frequently much smaller in size. This is because the Si—H bond is stronger and shorter than the Si—Si bond.

According to the degree of hydrogenation, the spectra (**Figure 1**) clearly illustrate the effects of the two hydrogen esteems on the amorphous silicon lattice. Only two peaks can be found in the red spectrum, and they are $2\theta_1 \simeq 28.32°$, and $2\theta_2 \simeq 47°$ in terms of diffraction angles. Further low peaks signaling the emergence of a desired trend in the material's behavior. The action of hydrogen causes an increase in crystallinity, and the size of the crystallites ($\simeq$ 71.5 pm) demonstrates that this increase is not consistent with polymorphous silicon.

4. FTIR analysis

The raw FTIR spectra were firstly corrected for incoherent and coherent reflections. In addition, the correction of the absorption of the film (a-Si: H) was obtained by subtracting the measured absorption of a bare part of the substrate. The observable peaks are due to the absorption states caused by the vibrations of various bonds and rely upon the encompassing environment which likewise doesn't have a comparative energy level. The vast number of peaks in the frequency region between 450 cm^{-1} and 1800 cm^{-1} is due to the fact that its existing conformation is exceptionally wealthy in distinctively situated atomic chains and furthermore their diverse sub-joins.

It has been recognized that an increase in the Si—H_2 concentrations leads to a weakening of the photovoltaic properties which subsequently leads to poor device performance. The above physical conditions are consistent with the following specific bond units: Si—H and Si—H_2, but we can rarely find Si—H_3 components. Infrared spectroscopy has been used to get the concentrations for the possible configurations $Si-Si_{4-m}H_m$ (m = 0 ... ,3) considered in the Tetrahedron model [32]. The resolve of the hydrogen content is determined through different processes like elastic recoil detection analysis (ERDA). For our situation, the hydrogen content was calculated by means of the infrared absorption range at the 640 cm^{-1} frequency addressing the wagging mode, just as the density of the $Si - H_x$ bonds and any remaining fragments as—$[SiO_{\ominus}]$—; that might be shaped inside or at the sample-substrate border. The quantitative process of the IR modes is based on the accurate measurement of the hydrogen content through the oscillator strengths. The density N_{Si-H_x} (x = 1,2) is proportional to the integrated absorption strength of the given mode [33]:

$$N_{Si-H_x} = A_n \int \frac{\alpha_{eff}(\omega)}{\omega} d\omega = A_n\, I, \alpha_{eff} = \frac{A\ (\omega)}{d.\log_{10}(e)} \quad (1)$$

The effective absorption coefficient was calculated using Eq. (1), where d is the thickness of the thin film; the integral I is the absorption strength of each absorption peak, and A(ω) is the absorbance as a function of the series of permissible infrared frequencies ω. The function within the integral sign (Eq. (1)) is certainly based on the infrared frequency of the vibration of a particular set. The integral traverses the whole range of frequencies proving in any case the presence of such a gathering. The understood functional difference was presented in the form of a peak which can be displayed by a Gaussian function having an area A. The completely perfect fit allows calculating the area of such function, which is the integrated absorption strength I. The proportionality constant A_n show a discrepancy as the inverse of the oscillator strength, and it is independent of the total hydrogen concentration, and the integral is over the absorption band of interest. The matrix constant A_n has the dimension of a surface concentration. Some authors have highlighted some values, which are subsequently justified passing through numerous other works. For the wagging mode ($\simeq$ 640 cm^{-1}), it is $(2.1 \pm 0.2)\ 10^{19} cm^{-2}$, and for the low stretch mode ($\simeq$2000 cm^{-1}); it is $(9.0 \pm 1.0)\ 10^{19} cm^{-2}$. Lastly, the matrix constant identified with the high stretch mode ($\simeq$ 2100 cm^{-1}) has the worth $(2.2 \pm 0.2)\ 10^{20} cm^{-2}$ [34]. These matrix components quantitatively characterizing the microstructure were determined by estimating the beam yield of a resonant nuclear reaction of ^{15}N ions with hydrogen [35]. They are useful for numerically assessing the concentration of hydrogen and the densities associated to Si—H and $Si—H_2$ moieties, respectively, through the integral relationship referred to above (Eq. (1)). The density of the hydride bonds (Si—H and $Si—H_{\ominus}$) is calculated either by analyzing the FTIR absorbance spectrum or according to the infrared absorption coefficient by applying the above mentioned method. The hydrogen concentration is obtained by dividing the concentration of groups such asSi—H/$Si—H_{\ominus}$,$Si—H_{\ominus}$ and Si—H by the atomic concentration of crystalline silicon ($N_{c-Si} \approx 5 10^{22} cm^{-3}$):

$$C_H(at\%) = N_H/(N_H + N_{c-Si}) \quad (2)$$

Three distinctive wavenumbers ranges are recognized that compare to absorption modes of Si—H bondings: wagging modes around 640 cm^{-1}, bending modes around 860 cm^{-1} and stretching modes around 2000 $cm^{-1} \rightarrow$ 2100 cm^{-1}. The wagging mode is relative to the total hydrogen concentration [36], this confirms that the components of the array of clusters as well as the distributed hydrogen are equal under this mode.

We have tracked the aforementioned ways, the total hydrogen concentrations in the microstructure will inevitably increase with the level of hydrogenation gas. Accordingly, the C_H values for both films are $\simeq$ 16.18% (100 sccm) and $\simeq$ 33.96% (200 sccm).

4.1 Bending band

The infrared spectra for the hydride setups of a-Si: H films are centered on the range of bending modes and show all sub-modes (**Figures 2** and **3**).

The assessment of the infrared tops at the level of the absorption most extreme and at the level of the half-width shows critical contrasts coming chiefly from the hydrogenation. The subsequent peaks arrangement is decreased in size and on number as the hydrogenation stream parts (100 sccm $\rightarrow$ 200 sccm). The peak showed up at the frequency 450 cm^{-1} (**Figure 2a**) moves towards 500 cm^{-1} (**Figure 2b**) with decrease of the most extreme absorption and of the half-width and shows a stretch mode relative to the vibrating Si—Si bond.

It also shows a critical decrease of the deformities, as the dangling bonds and in contrast the prompt expansion in the number of Si—H dipoles. The volume fraction involved by Si—H dipole is lower than that of a homopolar Si—Si dipole seen that the Si—H bond is shorter than that of Si—Si, which implies that the number of Si—Si bonds establishes absorption field which widens its range by means of their void/a − Si arrangements. The chains $(Si—H_{\ominus})_n$ have a lot of similar bonds, and the vibrations can be conjugated, which leads to infrared absorptions at characteristic frequencies linked to the vibrations of chemical clusters. The doublet appeared between 800 cm^{-1} and 890 cm^{-1} indicating the presence of vibrating $(Si—H_{\ominus})_n$ chains in scissors mode. Moreover, the doublet showed up between 950 cm^{-1} and 1000 cm^{-1} (**Figure 2a**) indicating a degenerate mode connected to the presence of isolated $Si—H_{\ominus}$ hydrides in transverse vibration (Scissoring and Rocking). Consequently, it was reconstructed to make a single peak for a similar part of hydride in scissors mode vibration (**Figure 2b**). This decrease in the instability of the blended compositions is due to the buildup of the substance by hydrogen. In the frequency interval [600 cm^{-1}, 700 cm^{-1}] (**Figure 2**), two convoluted peaks appear with the exception of the size and the absorption extremum which undergoes a slight variation due to the concentration of hydrogen bound. It appears that a degenerate out-of-plane longitudinal vibration of the wagging/twist type is established and persists for the $Si—H_{\ominus}$ fragments [13]. We conclude that bound hydrogen reduces the degree of degeneration of vibrational dynamic states of hydrides. Conversely, in the frequency margin [900–1000 cm^{-1}] (**Figure 2a**) two convoluted peaks reflecting the breaking of the chains $(Si—H_{\ominus})_n$ for turn out to be isolated (**Figure 2b**). These two degenerate modes combine to be in a single high extremum absorption mode reflecting the decrease in vibrational instability from the fragments of the polymeric hydrides. Therefore, increasing the hydrogenated configurations then reorganizes the microstructure by reducing the modes of vibrational degeneration due to the $Si—H_{\ominus}$ units. In contrast, a structure with low hydrogen content removes the vibrational degeneration of the hydrogenated fragments, reflecting fractional levels of apparent mixed-phase energies that may act as an induced dipole.

Likewise, for a flow of 200 sccm, **Figure 3b** show an extreme reduction of IR peaks resulting from the formation of Si—H dipoles (**Figure 3a**). Here again, the role of hydrogen seems to passivate the unordered a − Si/Void arrangements by in turn

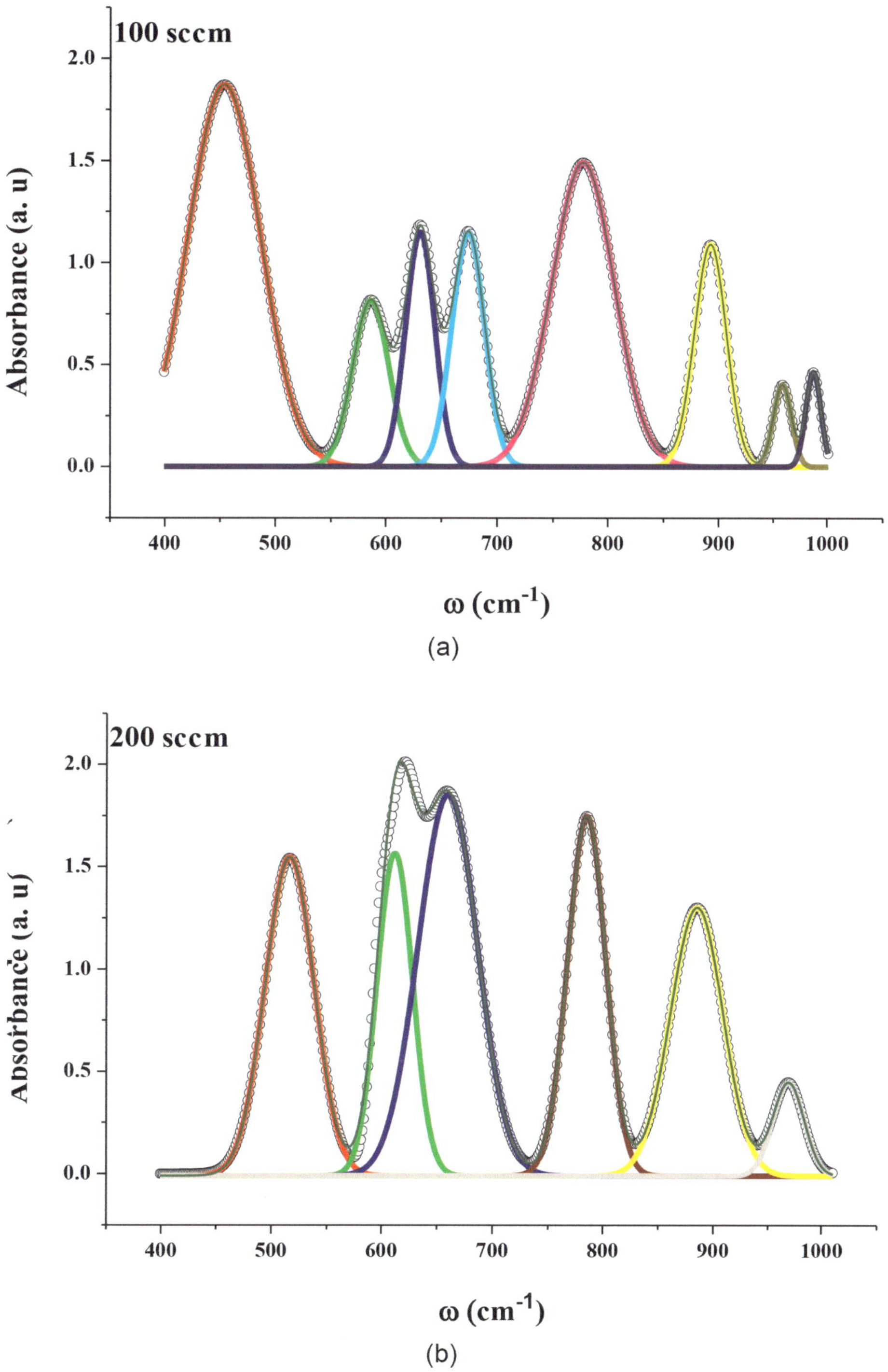

Figure 2.
Deconvolution of bending vibration modes in the range [400–1000] cm^{-1} for two consecutive hydrogenation flow of 100 sccm (a) and 200 sccm (b).

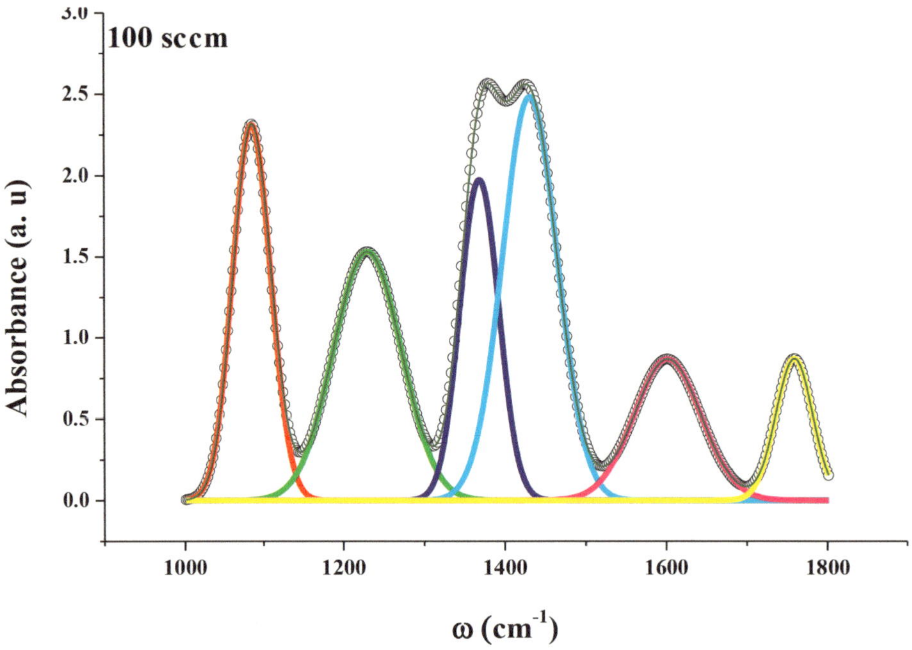

(a)

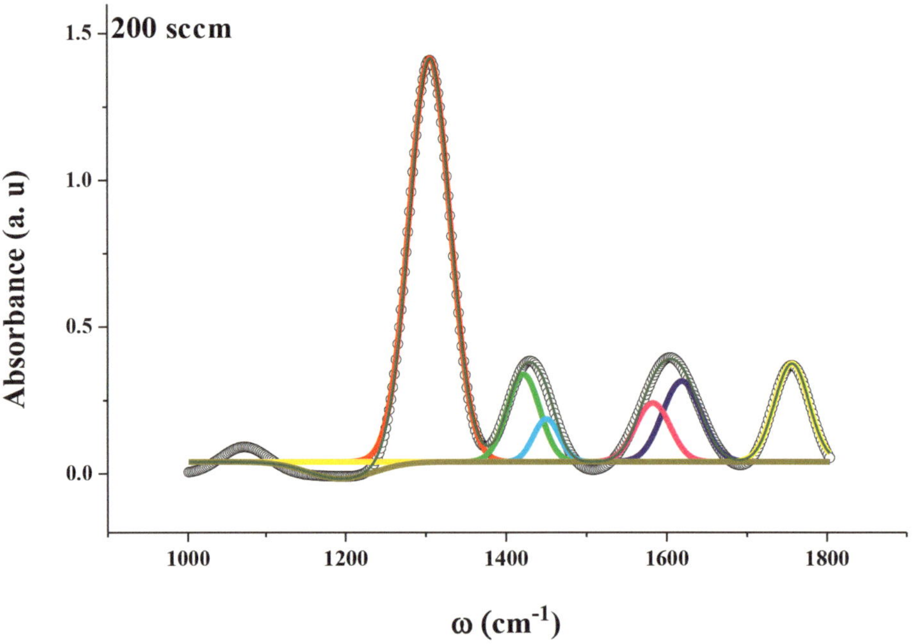

(b)

Figure 3.
Deconvolution of vibration modes in the range [1000–1800] cm^{-1} for two consecutive hydrogenation flow of 100 sccm (a) and 200 sccm (b).

increasing the characteristic gap (Eg $\simeq$ 4 eV), which has already been demonstrated by simulating the data via the Tauc- Lorentz model (**Figure 4**).

The peaks (**Figure 3a**) appeared near 1208cm^{-1}, 1400 cm^{-1} and 1800 cm^{-1} showing the formation of Si—O bonds by developing clusters $-[Si—O_x]-$ in which the non-passivated part adheres to the vitreous substrate. These boundary configurations were further marked by the XRD spectrum resulting from oxidation at the $a-Si:H$/glass interface following locally required areas. Hydrogen reduces absorption and widens the characteristic range, showing only a peak near 1250 cm^{-1} refers to a rolling absorption center (**Figure 3b**). The peak centered in the vicinity of 1600 cm^{-1} relates to a low hydrogenation Void/Si O_2 composition, but when more bound hydrogen is added, such a composition has become bilateral and is less energetic. Hence, the identical fragments (O=Si $-$ O,or O $-$ SiH_2 $-$ O) are in fact crossed and have not yet had a dynamic conservative mode [37]. Infrared absorption spectra relating to bending mode show the reduction of the microstructure of hydrogenated amorphous silicon via the increase in the hydrogen flow which pushes the multiplication of Si—H and Si—$H_\ominus$ bonds. This multiplication rearranges the tetrahedral conformations by replacing a silicon atom with a hydrogen atom, and thus an extreme reduction in the density of the dangling bonds. On the other hand, the absorption rate is more increasing when the structure is purely amorphous formed by silicon atoms. The gain of one hydrogen atom in the new microstructure lowers the absorption power. **Figure 3b** clearly shows the minimization of the number of absorption peaks when switching to a more enhanced hydrogenation flow. Previous work has shown that hydrogen can push the atomic grid back to a more ordered state through enhanced minimization of the built-in void and thus through the change in electrical properties caused by the entrainment of heteropolar bonds [38].

4.2 Stretching band

While the aggregate sum of hydrogen is regularly assessed from the integrated intensity of the wagging modes, the stretching modes contain more data about the microstructure and the voids fraction in a-Si: H. There is general agreement that there are at any rate two stretching modes. To resolve the densities of the hydrogenated Si—H-type configurations and that of Si—$H_\ominus$ -type in the bulk environment, the stretch band deconvolution measure is performed by two Gaussians focused separately at 2000cm^{-1} and 2100 cm^{-1}. Given the 7059 glass substrate used to deposit the film, it is possible to have connections between the mass of the film and the boundary particles of the glass substrate. Hence, to find out the intensity of these atomic configurations responsible for this interconnection that locally brings a new aspect to the film such as a nanocrystal or a microcrystal, it is necessary to decompose the spectral portion by Gaussians centered on all the Eigen frequencies, and thus calculate the required intensities. Accordingly, at the substrate interface, the limit bunches act like a little part of the film perceiving the monocrystalline or microcrystalline profile. There, this perspective can react with absorption modes reaching out around 1890 cm^{-1}–1975 cm^{-1} (ELSM). In the mid-frequency range located at 2030 cm^{-1}$-$2040 cm^{-1} (MSM), the limit bunches basically causes absorption groups via platelet surfaces (chain shut by the succession of patterns: $[-(Si-SiH)_n-]$ and most likely reflect restricting fragments in the substrate-film interface. What's more, it ought to be noticed that there are different groups beginning in the frequency positions 2083 cm^{-1} and 2137 cm^{-1}, and

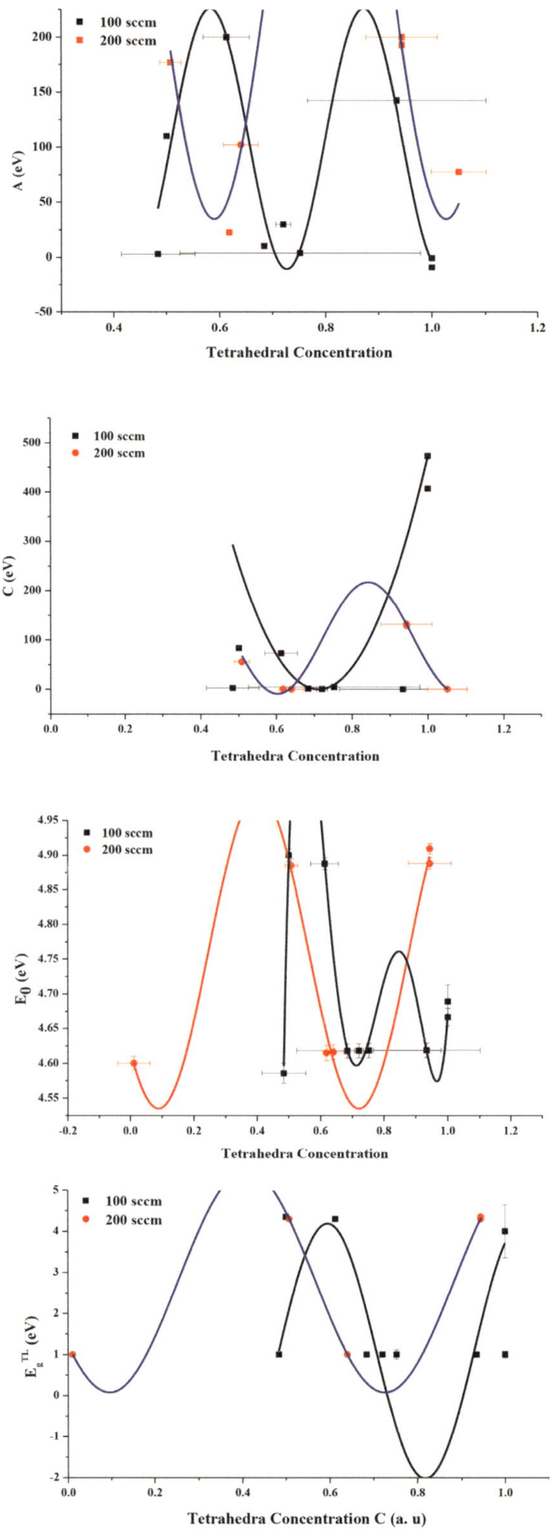

Figure 4.
Tauc-Lorentz parameters as a function of tetrahedra concentration: A, C, E_o *and* E_g *respectively.*

$$V_{void} \approx S_{void}{}^{3/2} \approx C_{HSM}{}^{3/2}$$

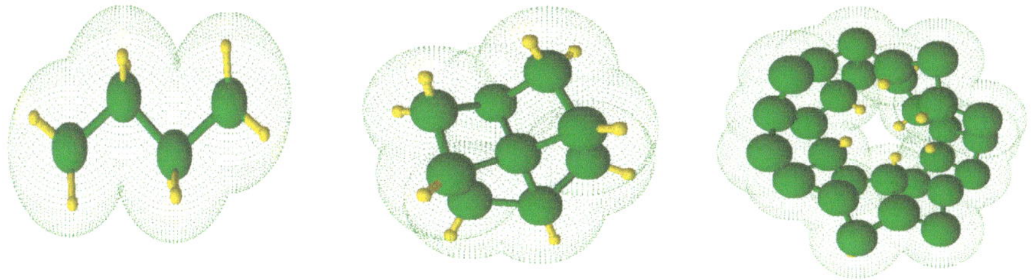

Figure 5.
Growth of Si-H_x bond fragments until the formation of fully passivated spherical microcavities: Linear chain of low density in Si − H *and* Si − H_2 *(a), Atomic surface with* Si − H_2 *bonds at the outer limits (b) And a microvoid with internal walls of* Si − H_2 *bonds (c).*

ELSM ~1890–1970 cm^{-1}	$(\mathbf{SiH})_n$ located near the interface substrate-film with mutual hydride dipole-dipole interactions: bond centered H **Si** ≡ **Si** − **H** ... − **Si** ≡ **Si**
LSM ~2000 cm^{-1}	MHs (SiH) restricted in the internal tissue of a-Si: H and form divacancies.
MSM ~2030–2040 cm^{-1}	Platelet surfaces: A chain shut by the succession of patterns $-[(Si - SiH)_n]-$, located in the bulk mass.
HSM ~2099±2 cm^{-1}	DHs (SiH_2) limited in the bulk mass of a-Si: H, unscreened configurations grafted into the internal walls of nanovoids.
NHSM ~2083–2103–2137 cm^{-1}	Micro-surfaces of $SiH_x (x = 1, 2, 3)$ with less dense hydrogenated configurations at high level of crystallinity, recognized in crystal grain boundaries surrounded by mass.

Table 1.
Infrared absorption modes reflecting all potential arrangements in hydrogenated amorphous silicon.

which exhibit the presence of the Si − H_x configurations (x = 1, 2 and 3). These groups so-called NHSM were set apart on little glasslike regions situated at the bulk material grains of the film [37, 38]. We consider that these groups are because of hydrogenated crystal surfaces, which are basically present in the less thick μc-Si: H with high crystallinity (see **Figure 5**). Besides, NHSMs modes show the presence of smooth surfaces containing accretions of mono-, di, and tri-hydride [39], which thus accentuate the edges of voids drawn with valuable patterns in the folds of the thin film (**Table 1**) [38]. This outcome is in concurrence with the investigation of the stretch modes in the frequency range 1900–2250 cm^{-1} (**Figure 6**). It relates to HSM mode a hydrogen bonds to the inner surface of the microcavities [4].

To envisage the change of the density of the hydrogenated setups in the range of the stretching mode frequencies and in the case of higher hydrogen stream, it is certain that they change as per them dimensions and the light absorption rate (**Figure 7**).

$$N_{str} \approx \sum_{p=0}^{3} \ell^{p} \omega_{p}^{p+3} \quad (3)$$

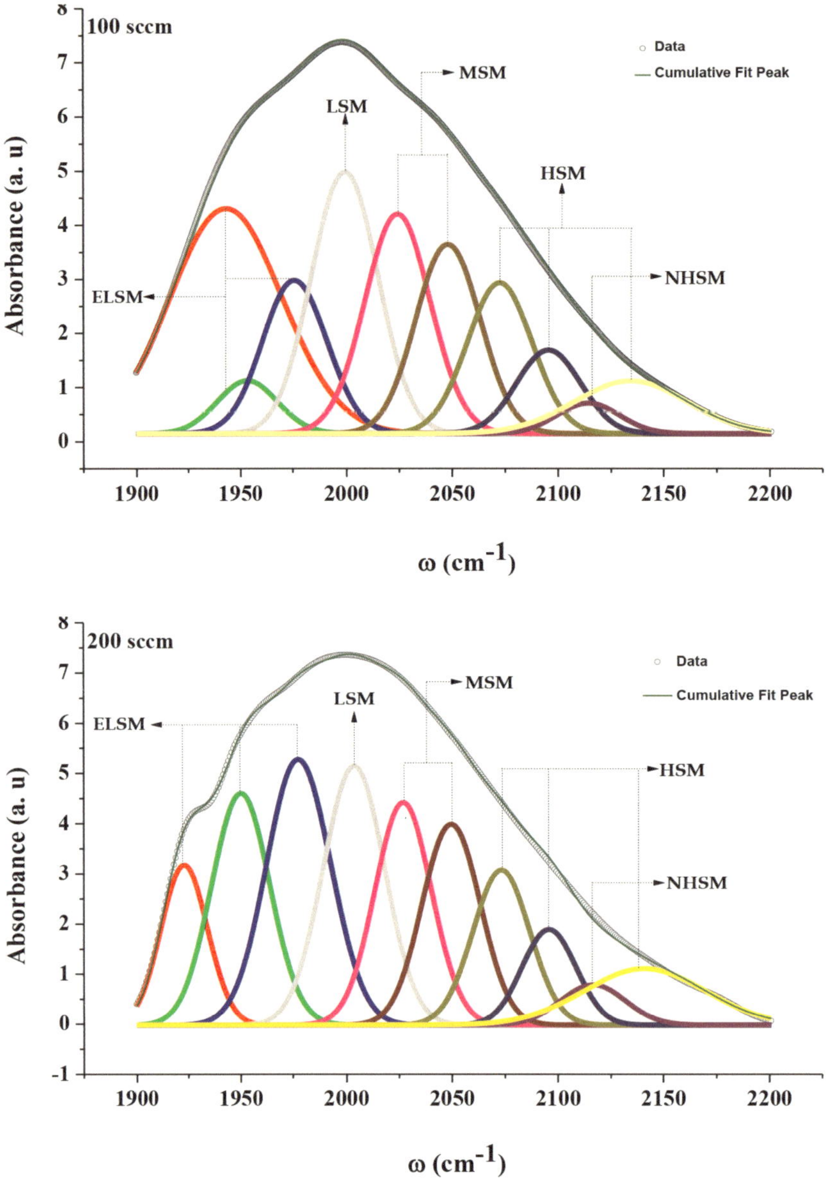

Figure 6.
Frequency deconvolution of the stretching mode for two hydrogenation flow, 100 sccm and 200 sccm, respectively.

ℓ^p is a dimensional measurement of the gathering vibrating at the frequencyω_p. Eq. (2) was expected through the experimental curve (**Figure 7**) to attempt to clarify how the density of $Si - H_x$ (x = 1, 2, 3) formations increments. We assessed that these advance because of the kind of coupling made in the microstructure. The components of the molded chains were entered into the equation, and this is referred to as the power p, whether it is open or shut.

In the case that they are open, the proven density is minor, and when these congregations are shut, the density is more contrasted with the situation where the

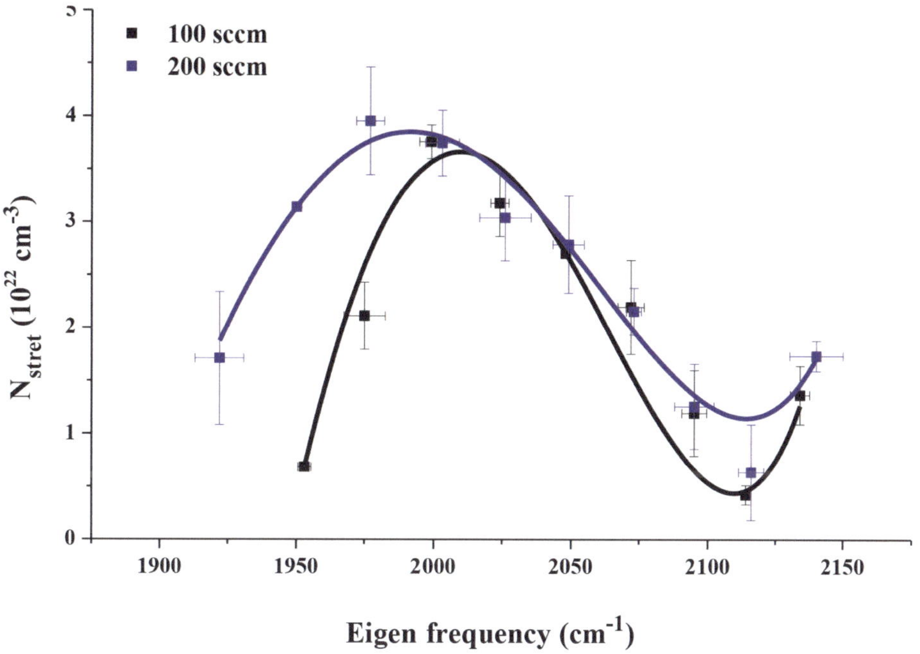

Figure 7.
Evolution of the density of monohydride and dihydride configuration as a function of all Eigen frequency.

chains are open. With respect to the related frequencyω_p, it considers by its worth the sort of arrangements, i.e., Si—H,Si—$H_\ominus$ or $(Si—H_\ominus)_n$. As well, the dimension ℓ^p addresses the normal length of Si—$H_\ominus$ and Si—H groups in the amorphous silicon lattice. For p = 2 (separately p = 3), ℓ^p mirrors the typical region and the normal size involved by the restructured round Nano-Configurations [40].

This semi noticed depiction consolidates dimensional boundaries such as the changes in the vacancy size and the diameter of nanovoids; that advance the development of hydrogenated microstructures [41]. Nevertheless, absorption sub-modes mirroring the presence of hydrogenated bunches close to the limit substrate-film or in bulk mass; concede different frequencies not the same as those referred to for the stretching mode as a whole, chiefly LSM and HSM. The frequency shift $\Delta\omega_{SM}$ restricted by the edge of a characteristic frequency ω_p and a particularly unknown frequency $\omega_{p'}$ ($\simeq \omega_p \pm \Delta\omega_p$) related with such bunches yet to be noticed gives, among others, the bulk positions of monohydrides Si—H.

It is likewise connected to the nanoenvironnement of Si—H configurations in stretch mode. This frequency shift between such two stretching sub-modes is given by the following relation [38]:

$$\Delta\omega_p \approx -\frac{10^{-4}}{24\pi^2 c^2 m_p \omega_{0,p} \varepsilon_0} N_p q_{p,e}^2 \tag{4}$$

The frequency $\omega_{0,p}$ assigns the p-eigen frequency; it is taken as the first limit of the frequency shift, m_p is the mass of a Si—H dipole in the hydrogenated packs dwelling in a micro-cavity of volume V, N_p is the density of the dipoles (Si—H or Si—Si) in the screened phase at the p-mode, and $q_{p,e}$ is the effective charge of p-mode. In this way, for any shift that happens, it is in every case a lot of lower than the frequency ω_p (i.e.$\frac{\Delta\omega_p}{\omega_p} \ll 1$). In this manner, we can develop Eq. (2) in arranged power series:

$$N_{str} = \sum_{p=0}^{3} \ell^p (\omega_p + \Delta\omega_p)^{p+3} \approx \sum_{p=0}^{3} \ell^p \omega_p^{p+3} \left(1 + \frac{\Delta\omega_p}{\omega_p}\right)^{p+3}$$
$$\approx \sum_{p=0}^{3} \ell^p \omega_p^{p+3} \left(1 + (p+3)\frac{\Delta\omega_p}{\omega_p}\right) \approx \sum_{P=0}^{3} \left(\ell^p \omega_p^{p+3} + p\, \ell^p \omega_p^{p+2} \Delta\omega_p + 3\Delta\omega_p \ell^p \omega_p^{p+2}\right) \quad (5)$$

If p traverses these values granted, then, the density of the hydrogenated phase ($Si—II_x$) in the stretching mode is given by the sum above showing a set of eleven terms, each of which contributes to a certain density of sequences of dipoles located at well-suited sites. The only difference between all stretch sub-modes is the area of ELSM (Extreme Low Stretching) modes around the following position ~1943 cm^{-1}, ~1950 cm^{-1}, ~1975 cm^{-1} (**Figure 6**). Even so, it can be seen that the relationship between the two batches of ELSMs may result from a change in a single material composition. Hence, this is due to a very low fraction of hydrogen bonded. Additionally, these absorption modes show a net frequency shift which appeared also with respect to the modes (LSM and HSM) reflecting the presence of configurations of the Si—H, $Si—H_{\ominus}$ or in like manner tri-hydride ($Si—H_{\oplus}$) type on crystalline surfaces credited to the limits of the vitreous grains in the mass. Such atomic configurations with hydrogen are relied upon to connect to extremely high hydride densities commonly responding through dipole-dipole interactions. **Figure 6** shows also that a more noteworthy extent of structures having the Si—H bond shows up with the LSM absorption mode. While the density way of the hydrogenated combinations goes through an upward sunken change focused on edges of 2100 cm^{-1}, this exhibits that the $Si—H_{\ominus}$ proportion is low contrasted with Si—H. This outcome was seen through the upgraded combinations affirmed by the reproduction of the SE results (**Figure 8**).

The variation of hydrogen content as a function of the Eigen frequencies of the stretching mode follows a polynomic function. The error which follows the end of the adjustment is minimal, and the data regression ends with a good degree of fit (R^2 = 0.90).

$$C_H(\%) \simeq \sum_{i}^{m} a_i \omega_j^{\,i} \quad (6)$$

According to the order of each of the multiplying coefficients for the j-frequency, we note that they sometimes have an approximate weighting of the lengths of the dissimilar bonds, and at other times the extent of the spherical microcavities of surface S which assumes the size of the voids recreated at a concentration more prominent than 14%. We suppose the factor a_1 as the approximate length of the chains $(Si—H)_n$. Then, a_2 is a surface and can also be a closed chain by weak extended $Si—H_x$ fragments. After that, a_3 is a volume, so we can envisage that there is a microvoid configured by adjusting a small offset of the surface of the chains $(Si—H_x)_n$ below the ω_3-frequency. We will set up the reliance of the multiplier coefficients on the cross part of the microcavities and along these lines with its mean degree.

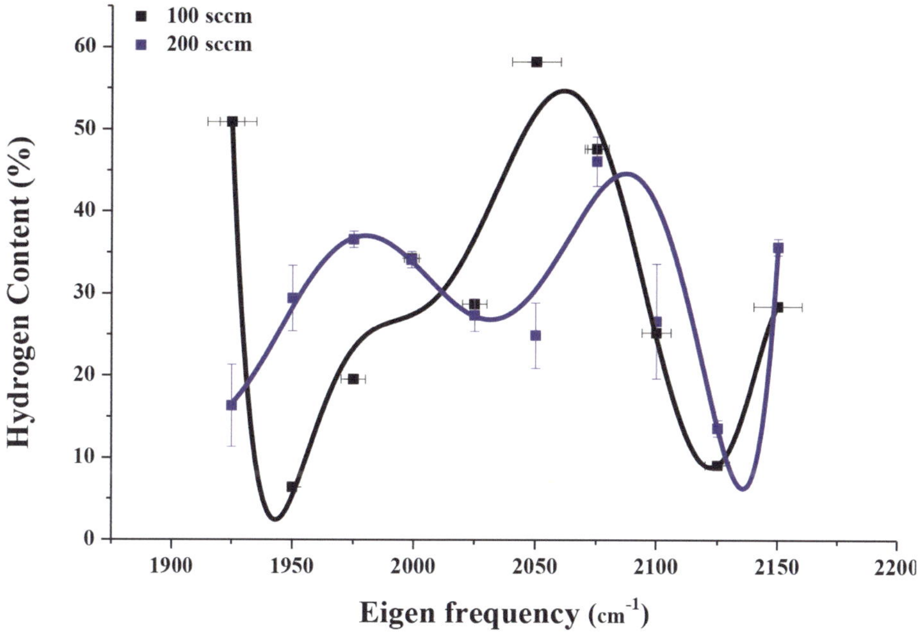

Figure 8.
Hydrogen content versus the Eigen frequencies of the stretching mode.

This outcome is in exceptionally legitimized concurrence with the reasonable mainstays of the ECMR model proposed by Drabold et al. [42]. Although the relation that considers the change in hydrogen concentration as part of the Eigen frequency of stretching mode, it is composed roughly as a component of the leveled surfaces with a given cross over degree:

$$C_H(\%) \simeq \sum_{n=1}^{N} s^{\frac{n}{2}}\, \omega^n \qquad (7)$$

This empirical equation is proportional to the comparing frequency of the infrared vibration of Si—H_x-type composition (x = 1, 2). Then, an expansion in hydrogen concentration permits the distinctive complex gatherings to bend or shift in the permitted directions, trailed by a brief change in the angles between the diverse substituents of the first framed straight series. As per a few works did for over 40 years, this additionally implies that the overall concentration is controlled by the statistics of the infrared vibration frequencies of the trademark clusters of n-request surfaces that are framed by the hydrogenated groups that are presently arrangement. Subsequently, the level of hydrogenation is the operating factor liable for the arrangement of spherical nanovoids in the overall matrix. Beyond 14% in hydrogen content which does not surpass a restriction of 23%, the multiplication coefficient a_3 is equivalent to $s^{3/2}$. This limit is consistent when we have it in the state where the absolute bound hydrogen concentration accurately surpasses 14%. It identifies with the measure of the volume of void shaped by the hydrogen holding chains, their region, and the level of hydrogen in the HSM mode as follows [4]:

$$V_{void} \approx S_{void}^{3/2} \approx C_{HSM}^{3/2} \qquad (8)$$

The all setups considered show the process of advancement of the hydrogenated arrangements from a state where the density of **Si − H** bonds is very low (**Figure 5a**), and until partial saturation by the formation of two-dimensional micro-chains mainly made up of slight zones (**Figure 5b**). The increment in hydrogen combined with silicon atoms enhances the creation of $(\mathbf{Si - H_2})_n$ chains which thusly start to squirm under the additional forces of electrodynamic states and eventually form vacuoles with embedded internal walls of $\mathbf{Si - H_2}$ fragments (**Figure 5c**).

5. Ellipsometry measurements

An optical model is addressed as a function of the refractive index and the thickness of the thin film (N, d). Knowing the refractive index of the substrate and its thickness, those of the thin film can be determined by treatment the ellipsometric data (ψ,Δ) by means of suitable adjustment model. Optical angles are interrelated by Snell's laws as a light shaft passes starting with one layer then onto the next. A very important fact, in the event that the absorption's film is low, this compares to a low refraction angle, which causes optical interference by a few reflections of light from the film's surface. In particular, there is a superposition of the interferometric fringes in a destructive manner. In this case, the total intensity of the reflected light becomes smaller and smaller towards high photon energies because of the surface properties including roughness. This explains the fact of the wide oscillations in the spectrum of the optical phase shift Cos(Δ) between the parts in the (p, s) plane of the light reflected ($r_{p,s}$) from the surface. To maximize the amount of reflected light, the incidence angle θ_0 should surpass the limited angle of refraction.

However, part of the incident light undergoes a surface scattering phenomenon depending on the roughness of the latter. Along these lines, the use of an incidence angle around 70°∼80° makes it possible to recover the minimum fractions of the reflected light absorbed by the surface and to avoid the light scattering by the effect of roughness [43–47]. The BEMA model shows more exact and nitty gritty outcomes than those of Maxwell Garnet (MG), Lorentz-Lorenz (LL), and provides the best fit in the analysis of surface roughness layers [48].

The countless materials compositions at last do not acknowledge the smallest change in their arrangement with one another (type or amount). The SE data never shows an arrangement such as Si_3H/Si_2H_2, and the reproduction sequence never approved such configuration during the relapse of the conceivable proposed compositions, which shows that the film does not contain any setups framed by monohydrides and dihydrides chains stable.

These outcomes have been proven after 6,000 emphases. Film thicknesses were determined by SE measurement. Appropriately, for a hydrogenation gas stream F_H equal to 100 sccm, the current film has a thickness around 213 nm; the second has 183 nm for F_H equal to 200 sccm.

Brugemman's approximation show nine phase materials, each with its own dielectric response and that make up the whole sample (see **Figure 9**). The dielectric response for every mixed phase was acquired through setback of SE measurements. So as to get and investigate its optical properties, we present our optical examinations dependent on the SE, just as the determination of dielectric function. In fact, hydrogen has numerous impacts in silicon matrix; such as, the remaking of frail Si—Si bonds, the nearby densification of the amorphous network because of shorter length

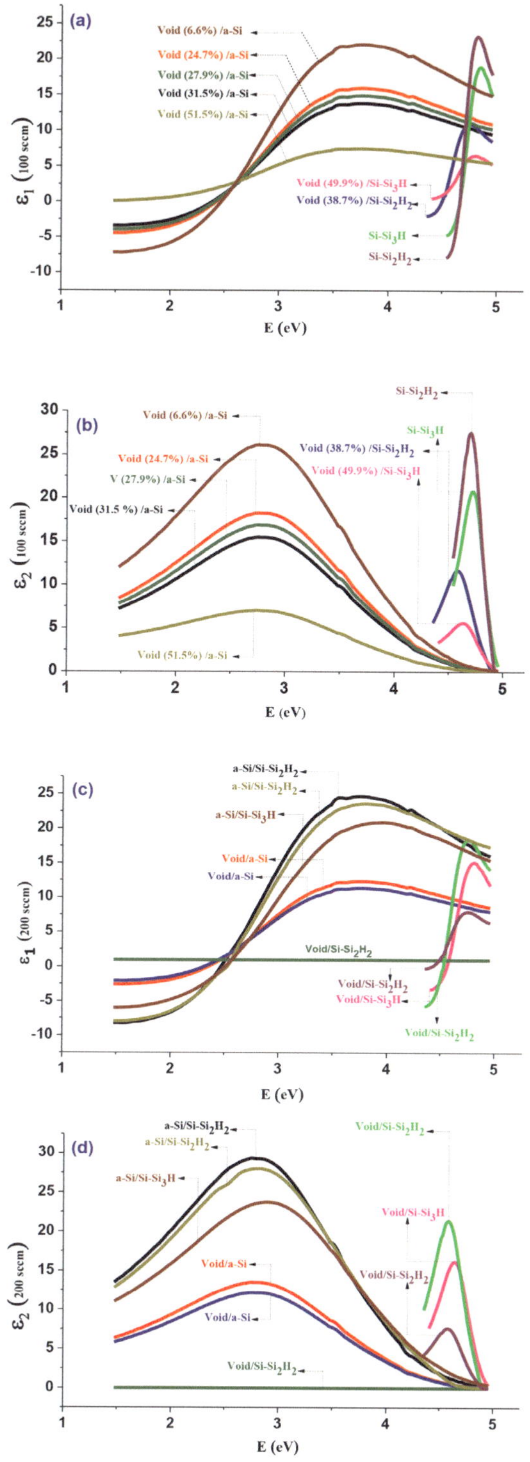

Figure 9.
Diagram of ϵ_1, ϵ_2 for all mixed phase in hydrogenated amorphous silicon: (a) and (b) relate respectively to the real and imaginary parts for a 100 sccm hydrogenation flow, (c) and (d) for the two dielectric parts respectively for a flow of 200 sccm.

of the Si—H bond in contrast with the Si—Si bond, the bringing down of network cross-linking, since of an improvement of void growth, and an expansion of the density of states in the conduction band because of the antibonding states of the Si—H bond. The void used as an explanatory factor of the optical properties of hydrogenated amorphous silicon has not been threw in the folds of the atomic network with proportions known by a particular procedure, but is rather present in its impressions at each corner of the film, it is a very important characteristic of the hydrogenated amorphous silicon, and we have shown the extent of its effect on the optoelectronics properties.

5.1 Dielectric function

The dielectric spectra were defined by the system of Eq. (9) which is the theoretical foundation of the Bruggeman model [2]:

$$\begin{cases} \sum_i v_i \left(\epsilon_i - \epsilon / \epsilon_i + 2\epsilon \right) = 0 \\ \text{avec} \\ \sum_i v_i = 1 \end{cases} \tag{9}$$

For the various mixed compositions of each thin film, the diagrams of the real part ϵ_1 and imaginary part ϵ_2 of the dielectric function were calculated as functions of photon energy (**Figure 9**). The parameters ϵ_i and v_i separately address the dielectric functions; the volume part of the nth component, and ϵ is the measured effective dielectric function of the film. Therefore, in the case of a random distribution of hydrogen bonds, each mixed composition has a volume fraction f_m given by the following relative expressions:

$$\begin{cases} f_m = \dfrac{P_m(x) V_m}{\sum_m P_m(x) V_m} \\ \varrho = \sum_m \varrho_m v_m \end{cases} \tag{10}$$

The mass density of the whole film ϱ is expressed as a function of the elementary densities ϱ_m of each tetrahedral composition. In essence, the volume that each material occupies relies on the number of bonds formed with the tetrahedral silicon, with the material compositions inside each having a sufficient opportunity to form based on the atomic percentage of hydrogen (Si—H or Si—Si). Every composition, thus, has a distinct density. The flow of hydrogen affects how many bonds ($Si—H_x$) there are. As a result, the optical index will change as a result of the qualities it has developed into in order to detect the impact caused by the disorder in the material and the different composition ($V/Si_{4-m}H_m$).

The dominant proportion of hydrogenated clusters can be calibrated according to the ratio $p_{1/2}$ = (N_{Si-H}/N_{Si-H_2}). For the situation that the hydrogen concentration is less than 14%, implying that $p_{1/2}$ is more than 16, the proportion of monohydride configurations dominates on the silicon network. Interestingly, the dihydrides dominate the folds of the microstructure, which thus create a low rate of stable divacancy as expected by numerous scientific works; this is the thing that we got through the regression of ellipsometric data.

The SE measurements provided very precise values that could be linked to some facts.

The variational shapes depend on the distribution of the nanovoids and on the extent of the change in its size as a function of the hydrogen content. Liu et al [49] show that the highest refractive index value indicates a very dense microstructure, which is the case of the following hydride composition: $a - Si/Si - Si_2H_2$ or $a - Si/Si - Si_3H$. In the Near infrared, the refractive index diminishes with expanding dose. This is a contention and proof of hydrogen passivation of the sample. If the index increments in UV and diminishes in infrared shafts, around then, this mirrors that the substance is passivated by hydrogen [13]. The density of the different bonds and the change in the voids size are the first makes driving unraveling the optical spectra. Just as, the dielectric functions are acquired for various individual coordinated amorphous silicon, which ought to be available in the movies as $Si - Si_{4-m}H_m$ joined by some fraction of voids. In the event that a layer is completely made out of hydrogenated amorphous silicon whatever his state of hybridization; at that point the level of control of the void is zero (the case of $Si - Si_2H_2$ or $Si - Si_3H$). The curves identifying with the structure Void/a-Si have the equivalent nearly variety profile along the energy extend. However, the distinction at a noteworthy quality (most extreme and least) is clearly because of the extents of the arrangements $Si—Si_{\oplus}$. The maxima of variation as indicated by the energy of these configurations diminished in sufficiency when the level of the void increments. It has the most reduced amplitude ($\simeq$ 7.48 SI), and which relates to the most elevated level of void ($\simeq$ 52%). Oppositely, the least worth ($\simeq$ 7%) is related with a similar mixture having amplitude of the request for $\simeq$ 22.104 SI.

The dielectric response of $V/Si_{4-m}H_m$ ($m \neq [0, 3[$) compositions showed up at exceptionally high energy, unequivocally past 4.35 eV, they are gotten by supplanting a Si—Si bond by Si—H in the tetrahedral configuration $Si - Si_4$. By disregarding the compositions of $a - Si$, it is seen that the profiles are requested then again regarding the presence of the setup having $Si - H_2$ and that conveying the $Si - H$. This is additionally because of the strength of $Si - H$ bonds, which need more energy ($4.5 - 5$ eV) compared with $Si - Si$ bonds (**Figure 9b** and **d**). The substitution of more grounded ($Si - Si$) bonds diminished the maximum real and imaginary part of dielectric function. This could likewise be proof that the dispersion of hydrogenated clusters in the matrix of a-Si: H is exceptionally slight and far separated, isolated by silicon gatherings and various extents of the void [50]. The configuration $Si - Si_2H_2$ deprived of void has the most elevated estimation of dielectric amplitude (23.26 SI→ 4.82 eV, **Figure 9d**) trailed by $Si - Si_3H$ (18.95 SI → 4.84 eV, **Figure 9d**). As a result, the higher part of void is related to lower hydrogen content.

5.2 Mass density

Film density was obtained using the classical Clausius-Mossotti relation. It describes the dealing between the refraction index and film density. It is given by the following formula when it is 100% amorphous [51]:

$$\varrho = \frac{3\, m_{Si}\left(\frac{n_\infty^2 - 1}{n_\infty^2 + 1}\right)}{4\pi\left[2\alpha_{Si-Si} + \frac{C_H}{1 - C_H}\left(\alpha_{Si-H} - \frac{1}{2}\alpha_{Si-Si}\right)\right]} \tag{11}$$

n_∞ is the refractive index for the state where the wavelength is exceptionally high, and C_H is the total hydrogen concentration. Furthermore, similarly as the mass of the film should include the polarizability of each sort of bond present, here we have two kinds of bond: Si—H and Si—Si. The polarizability is equal to 1.36 10^{-24} cm^3 (respectively 1.96 10^{-24} cm^3).

In such manner, every material composition was viewed as a whole film stacked by bonds of the $Si - H_x$ (x = 1, 2) type, containing a specific concentration of hydrogen [52]. The previously mentioned equation was reapplied to compute the mass density ρ_m (m = 1… 9) of every one of them along with of percent occupancy versus void scattered in the film microstructure. We likewise thought to be in this setting that when we have a tetrahedral composition which doesn't contain Si—H bond, we can't hold the worth of the polarizability against zero. Instead, the equation must be taken into account counts as being equal to zero, as is the case with bonds of the Si—Si type. From **Figure 10**, it can be seen that there is a slight decrement in certain values of mass density for some mixed phases compared to roughly in the film (100 sccm), then, these values are the ones that contain the highest percentage of hydrogen bonded to silicon, or rather, a huge number of Si – H bonds. This fact shows that the completely passivated composition occupies the smallest parts of the waist in the folds of the film. Again, the film with the highest hydrogenation level was set up to be the most flexible and the most elastic (less stressful microstructure). Consequently, the inclusion of hydrogen in the silicon matrix causes a relative change in the a-Si cross-section.

It has as well been shown that it is potential to ascertain the mass density of each blended configuration using the Clausius-Mossotti equation, and since we are attempting in this work to conclusively decide the degree of changes that can occur in the atomic structure, according to its developmental real properties, it is expected that all mixed phase have around a similar mass, and as we demonstrate that the two film contain nine blended configurations (**Figure 9**), the absolute mass of the film will be

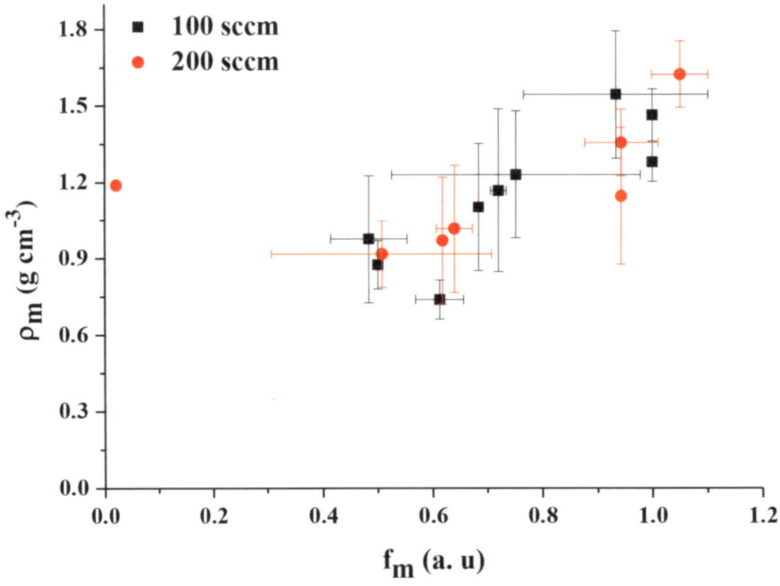

Figure 10.
Mass densities ρ_m for each tetrahedral configuration as a function of the volume fraction f_m.

the amount of every single sub-phase (see section 5.1 above). This critical prototypical assumes that in each case there is complete consistency of structural and discretionary properties.

5.3 Tauc-Lorentz Parameters

The counts were performed using Tauc-Lorentz model indicating the impact of void and hydrogen content on the bandgap of every hydride configurations. The gap of a-Si: H is changed by the disorder and with the hydrogen content, this implies that it is pretended by the all previously factors. According to Jellison et al. [53], the imaginary part ϵ_2 of the dielectric function can be written as:

$$\epsilon_2(E)^{TL} = \frac{1}{E}\frac{A\,E_0\,C\,(E - E_g)^2}{(E^2 - E_0^2)^2 + C^2E^2}\,H_t(E - E_g) \tag{12}$$

TL model contain four constants are treated as fitting parameters, H_t is the Heaviside distribution, E_0 is the energy of peak transition, C is the broadening term which is related to the disorder of the film and represents also a full width half maximum (FWHM) value of the normalized broadening Lorentz function, A is a factor related to the film density and proportional to the height of ϵ_2.

All data supporting the relative changes of the band gap energy were computed (**Figure 4**). Apparently the increment in the progress of hydrogen gas has encircled the band gap to the expansion. In fact, the hydrogen clearly influences the band gap only in the case while the a-Si: H is unsaturated in hydrogen (the case of void/a-Si) [54]. Instead, the gap retains a constant mean value independent of the concentration, assuming that it is fully saturated ($Si - Si_2H_2$). We once noticed that the gap increases when we have a completely hydrogen-free setup, again, if we had the opposite case.

The original reason why the gap differs in a monotonous manner could be the percentage of the void, that is, its size which affects the expansion of the internal range. As we have mentioned, an increase in the optical gap can result not only from an increase in the hydrogen concentration, but, also from the decrease of void ratio for each configuration [40].

Figure 4 shows this fact that the band gap reaches ~4 eV when the tetrahedral configuration is fully passivated. Since the distortion of Si—Si bonds in hydrogenated amorphous silicon is random, it results in the matrix of a-Si: H several possible values of the gap. Instead, a passivated bond removes a valence S-band in the material leaving valences formed primarily by deeper P-bands. This impact causes a critical broadening of the gap which can arrive at values more noteworthy than 3 eV. Moreover, it has been indisputably affirmed that the gap expansions with respect to the bound hydrogen content, that is, in entirely passivated tetrahedral configurations, as well, to the chance of bond among silicon and oxygen (= Si = O) invading from the substrate areas which improves this fact [55].

It can be concluded that the relevant hydrogen concentration has been changed, and this also brings a change in the thickness of the intended composition. Smets et al. [40] have shown that voids induced anisotropic volumetric compressive stress in the a-Si network leading to higher values of band gap energy. The disorder parameter (C) was not affected much by the rate of hydrogen more than the percentage of void in each composition.

All optoelectronic parameters evolve in a manner that varies according to the factor $(C_t - C_{t0})^2$ who C_t is the concentration of a specific tetrahedra composition and C_{t0} is a threshold concentration. This fact has already been acknowledged by Likhachev et al. [56], and a similar variation was also noted for the refractive index. The differences seen regarding the Tauc-Lorentz parameters may have a more rationale accommodating the output of the three parameters that generate the overall variation of ϵ_2.

6. Conclusion

As a function of the hydrogen and void concentration in each material composition, this chapter focuses on the dielectric and vibrational aspect of hydrogenated amorphous silicon. For each of the passivated configurations, a further relationship was established between the hydrogen content and the void proportion in order to explain how each configuration's characteristics relate to vibrational frequencies. Additionally, in connection to the eigenfrequencies of the stretching mode, the relationship between the densities of the $Si - H_x$ $(x = 1, 2)$ bonds and their spatial dimensions was taken into consideration and characterized. It has been found that, at the necessary synthesis circumstances, the completely passivated structures change precisely in terms of the degree of void, indicating a high gap that can reach multiple eV.

Author details

Mounir Kassmi
University of Tunis El Manar and Energy Research and Technology Center, Hammam Lif, Tunisia

*Address all correspondence to: mounirkassmi60@gmail.com

References

[1] Konagai MJJ. Present status and future prospects of silicon thin-film solar cells. Japanese Journal of Applied Physics. 2011;**50**(3R):030001

[2] Morral AF, Cabarrocas PR, Clerc CJP. Structure and hydrogen content of polymorphous silicon thin films studied by spectroscopic ellipsometry and nuclear measurements. Physical Review B. 2004;**69**(12):125307

[3] Saha JK, Bahardoust B, Leong K, Gougam AB, Kherani NP, Zukotynski SJTSF. Spectroscopic ellipsometry studies on hydrogenated amorphous silicon thin films deposited using DC saddle field plasma enhanced chemical vapor deposition system. Thin Solid Films. 2011;**519**(9):2863-2866

[4] Smets A, Kessels W, Van de Sanden MJA. Vacancies and voids in hydrogenated amorphous silicon. Applied physics letters. 2003;**82**(10): 1547-1549

[5] Stuckelberger M, Biron R, Wyrsch N, Haug F-J, Ballif CJR. Progress in solar cells from hydrogenated amorphous silicon. Renewable and Sustainable Energy Reviews. 2017;**76**:1497-1523

[6] Joannopoulos JD, Lucovsky G. The physics of Hydrogenated Amorphous silicon II: Electronic and Vibrational properties. 2008

[7] Robertson JJJ. Deposition mechanism of hydrogenated amorphous silicon. Journal of Applied Physics. 2000;**87**(5): 2608-2617

[8] Carlson DE, Wronski CRJAPL. Amorphous silicon solar cell. Applied Physics Letters. 1976;**28**(11):671-673

[9] Yang J, Banerjee A, Guha SJAPL. Triple-junction amorphous silicon alloy solar cell with 14.6% initial and 13.0% stable conversion efficiencies. Applied Physics Letters. 1997;**70**(22):2975-2977

[10] Shah A et al. Thin-film silicon solar cell technology. Progress in photovoltaics: Research and applications. 2004;**12**(2–3):113-142

[11] Kołodziej AJO. Staebler-Wronski effect in amorphous silicon and its alloys. Opto-electronics review. 2004;**12**(1):21-32

[12] Shimizu TJJ. Staebler-Wronski effect in hydrogenated amorphous silicon and related alloy films. Japanese journal of applied physics. 2004;**43**(6R):3257

[13] Kassmi M, Samti R, Dimassi W, Amlouk MJJS. Optical properties of material compositions in hydrogenated amorphous silicon through the degree of passivation. Journal of Non-Crystalline Solids. 2021;**562**:120771

[14] Paudel D, Atta-Fynn R, Drabold DA, Elliott SR, Biswas PJPRB. Small-angle X-ray scattering in amorphous silicon: A computational study. Physical Review B. 2018;**97**(18):184202

[15] Haage T, Schmidt U, Fath H, Hess P, Schröder B, Oechsner HJJ. Density of glow discharge amorphous silicon films determined by spectroscopic ellipsometry. Journal of applied physics. 1994;**76**(8):4894-4896

[16] Carlson DEJ. Amorphous silicon solar cells. Applied Physics Letters. 1977; **24**(4):449-453

[17] Takeda T, Sano SJM. Amorphous silicon position sensor for telephone terminal. MRS Online Proceedings Library (OPL). 1988;**118**

[18] Cody G, Tiedje T, Abeles B, Brooks B, Goldstein YJPRL. Disorder and

the optical-absorption edge of hydrogenated amorphous silicon. Physical Review Letters. 1981;**47**(20): 1480

[19] Sokolov A, Shebanin A, Golikova O, Mezdrogina MJJ. Structural disorder and optical gap fluctuations in amorphous silicon. Journal of Physics: Condensed Matter. 1991;**3**(49):9887

[20] Yamasaki SJPMB. Optical absorption edge of hydrogenated amorphous silicon studied by photoacoustic spectroscopy. Philosophical Magazine B. 1987;**56**(1): 79-97

[21] Daouahi M, Othmane AB, Zellama K, Zeinert A, Essamet M, Bouchriha HJS. Effect of the hydrogen bonding and content on the opto-electronic properties of radiofrequency magnetron sputtered hydrogenated amorphous silicon films. Solid state communications. 2001;**120**(5-6):243-248

[22] Maley N, Lannin JJPRB. Influence of hydrogen on vibrational and optical properties of a-Si 1− x H x alloys. Physical Review B. 1987;**36**(2):1146

[23] Gupta S, Katiyar R, Morell G, Weisz S, Balberg IJA. The effect of hydrogen on the network disorder in hydrogenated amorphous silicon. Applied physics letters. 1999;**75**(18): 2803-2805

[24] Bertran E, Andujar J, Canillas A, Roch C, Serra J, Sardin GJT. Effects of deposition temperature on properties of rf glow discharge amorphous silicon thin films. Thin solid films. 1991;**205**(2): 140-145

[25] Müllerová J, Fischer M, Netrvalová M, Zeman M, Šutta PJOP. Influence of deposition temperature on amorphous structure of PECVD deposited a-Si: H thin films. Central European Journal of Physics. 2011;**9**(5): 1301-1308

[26] Steffens J, Rinder J, Hahn G, Terheiden BJJ. Correlation between the optical bandgap and the monohydride bond density of hydrogenated amorphous silicon. Journal of Non-Crystalline Solids: X. 2020;**5**:100044

[27] Sánchez P et al. Characterization of doped amorphous silicon thin films through the investigation of dopant elements by glow discharge spectrometry: A correlation of conductivity and bandgap energy measurements. International Journal of Molecular Sciences. 2011;**12**(4): 2200-2215

[28] de Lima Jr M, Freire F Jr, Marques FJB. Boron doping of hydrogenated amorphous silicon prepared by rf-co-sputtering. Brazilian Journal of Physics. 2002;**32**:379-382

[29] Biswas P, Atta-Fynn R, Elliott SRJPRB. Metadynamical approach to the generation of amorphous structures: The case of a-Si: H. Physical Review B. 2016;**93**(18):184202

[30] Johanson RE, Guenes M, Kasap SOJIPC. Devices, and Systems, 1/f noise in hydrogenated amorphous silicon–germanium alloys. Devices, and Systems, 1/f noise in hydrogenated amorphous silicon–germanium all. IEE Proceedings-Circuits, Devices and Systems. 2002;**149**(1):68-74

[31] Mahan A, Williamson D, Nelson B, Crandall RJSC. Small-angle X-ray scattering studies of microvoids in a-SiC: H and a-Si: H. Solar Cells. 1989;**27**(1-4): 465-476

[32] Mui K, Smith FJPRB. Optical dielectric function of hydrogenated

amorphous silicon: Tetrahedron model and experimental results. Physical Review B. 1988;**38**(15):10623

[33] Mei J, Chen H, Shen W, Dekkers HJJ. Optical properties and local bonding configurations of hydrogenated amorphous silicon nitride thin films. Journal of applied physics. 2006;**100**(7): 073516

[34] Alvarez HDS, Silva AR, Cioldin FH, Espindola LC, Diniz JAJ. Hydrogenated amorphous silicon films deposited by electron cyclotron resonance chemical vapor deposition at room temperature with different radio frequency chuck powers. Thin Solid Films,. 2019;**690**: 137534

[35] Langford A, Fleet M, Nelson B, Lanford W, Maley NJPRB. Infrared absorption strength and hydrogen content of hydrogenated amorphous silicon. Physical Review B. 1992; **45**(23):13367

[36] Ouwens JD, Schropp RJPRB. Hydrogen microstructure in hydrogenated amorphous silicon. Physical Review B. 1996;**54**(24):17759

[37] Smets A, Matsui T, Kondo MJAPL. Infrared analysis of the bulk silicon-hydrogen bonds as an optimization tool for high-rate deposition of microcrystalline silicon solar cells. Applied Physics Letters. 2008;**92**(3):033506

[38] Smets A, Van De Sanden MJP. Relation of the Si H stretching frequency to the nanostructural Si H bulk environment. Physical Review B. 2007; **76**(7):073202

[39] Burrows VJAPL, Chabal YJ , Higashi GS, Raghavachari K, Christman SB. Applied Physics Letters. 1988;**53**:998

[40] Smets AH et al. The relation between the bandgap and the anisotropic nature of hydrogenated amorphous silicon. IEEE Journal of Photovoltaics. 2012;**2**(2):94-98

[41] Sekimoto T, Matsumoto M, Sagara A, Hishida M, Terakawa AJJ. Changes in the vacancy size distribution induced by non-bonded hydrogens in hydrogenated amorphous silicon. Journal of Non-Crystalline Solids. 2016;**447**:207-211

[42] Biswas P, Atta-Fynn R, Drabold DAJPRB. Experimentally constrained molecular relaxation: The case of hydrogenated amorphous silicon. Physical Review B. 2007;**76**(12):125210

[43] Archer R, Gobeli GJJ, Solids CO. Measurement of oxygen adsorption on silicon by ellipsometry. Journal of Physics and Chemistry of Solids.1965; **26**(2):343-351

[44] Herzinger C, Johs B, McGahan W, Woollam JA, Paulson WJJ. Ellipsometric determination of optical constants for silicon and thermally grown silicon dioxide via a multi-sample, multi-wavelength, multi-angle investigation. Journal of Applied Physics. 1998;**83**(6): 3323-3336

[45] Wakagi M, Fujiwara H, Collins RJTSF. Real time spectroscopic ellipsometry for characterization of the crystallization of amorphous silicon by thermal annealing. Thin Solid Films. 1998;**313**:464-468

[46] Fujiwara H, Koh J, Lee Y, Wronski C, Collins RJM. Real time spectroscopic ellipsometry studies of the solid phase crystallization of amorphous silicon. MRS Online Proceedings Library (OPL). 1998;**507**

[47] Koh J, Fujiwara H, Lu Y, Wronski C, Collins RJT. Real time spectroscopic ellipsometry for characterization and optimization of amorphous silicon-based

solar cell structures. Thin solid films. 1998;**313**:469-473

[48] Fujiwara H, Koh J, Rovira P, Collins RJPRB. Assessment of effective-medium theories in the analysis of nucleation and microscopic surface roughness evolution for semiconductor thin films. Physical Review B. 2000; **61**(16):10832

[49] Liu W et al. Characterization of microvoids in thin hydrogenated amorphous silicon layers by spectroscopic ellipsometry and Fourier transform infrared spectroscopy. Scripta Materialia. 2015;**107**:50-53

[50] Timilsina R. Microstructure, vacancies and voids in hydrogenated amorphous silicon. Mississippi: The University of Southern; 2012

[51] Remeš Z, Vaněček M, Torres P, Kroll U, Mahan A, Crandall RJJ. Optical determination of the mass density of amorphous and microcrystalline silicon layers with different hydrogen contents. Journal of non-crystalline solids. 1998; **227**:876-879

[52] Shaik H, Sheik AS, Rachith S, Rao GMJ. Microstructure dependent opto-electronic properties of amorphous hydrogenated silicon thin films. Materials Today: Proceedings. 2018;**5**(1): 2527-2533

[53] Jellison G Jr, Modine FJAPL. Parameterization of the optical functions of amorphous materials in the interband region. Applied Physics Letters. 1996; **69**(3):371-373

[54] Legesse M, Nolan M, Fagas GJT. Revisiting the dependence of the optical and mobility gaps of hydrogenated amorphous silicon on hydrogen concentration. The Journal of Physical Chemistry C. 2013;**117**(45):23956-23963

[55] Hanyecz I, Budai J, Szilágyi E, Tóth ZJTSF. Characterization of pulsed laser deposited hydrogenated amorphous silicon films by spectroscopic ellipsometry. Thin Solid Films. 2011; **519**(9):2855-2858

[56] Likhachev DV, Malkova N, Poslavsky LJTSF. Modified Tauc–Lorentz dispersion model leading to a more accurate representation of absorption features below the bandgap. Thin Solid Films. 2015;**589**: 844-851